Robin D. Howat, Mathematics Adviser
Edward C. K. Mullan, Galashiels Academy
Ken Nisbet, Madras College, St Andrews, Fife
Doug Brown, St Anne's High School, Heaton Chapel, Stockport, Cheshire

with

W. Brodie,   D. Donald,   E. K. Henderson,   J. L. Hodge,   J. Hunter,   R. McKendrick,
A. G. Robertson,   J. A. Walker,   P. Whyte,   H. S. Wylie

MATHEMATICS IN ACTION GROUP

CONTENTS

BLACKIE·CHAMBERS

MATHEMATICS IN ACTION

Mathematics in Action Group

Blackie and Son Limited
Bishopbriggs
Glasgow G64 2NZ

7 Leicester Place
London WC2H 7BP

W & R Chambers Limited
43–45 Annandale Street
Edinburgh EH7 4AZ

Illustrated by VAP Publishing Services, Kidlington, Oxford
Cover photograph courtesy of Rutherford Appleton Laboratory, Oxfordshire.

British Library Cataloguing in Publication Data

Mathematics in Action
   Pupils' book 4B
   1. Mathematics——1961.
   I. Mathematics in Action Group
   510   QA39 2

   ISBN 0–216–91927–4
   ISBN 0–550–75730–9 (Chambers)

ISBNs	Blackie	Chambers
Pupils' Book 4B	0 216 91927 4	0 550 75730 9
Teacher's Book 4B	0 216 91928 2	0 550 75731 7
Answer Book 4B	0 216 91929 0	0 550 75732 5
Assessment 4B	0 216 91930 4	0 550 75733 3

Printed in Great Britain by Scotprint Ltd, Musselburgh, Scotland

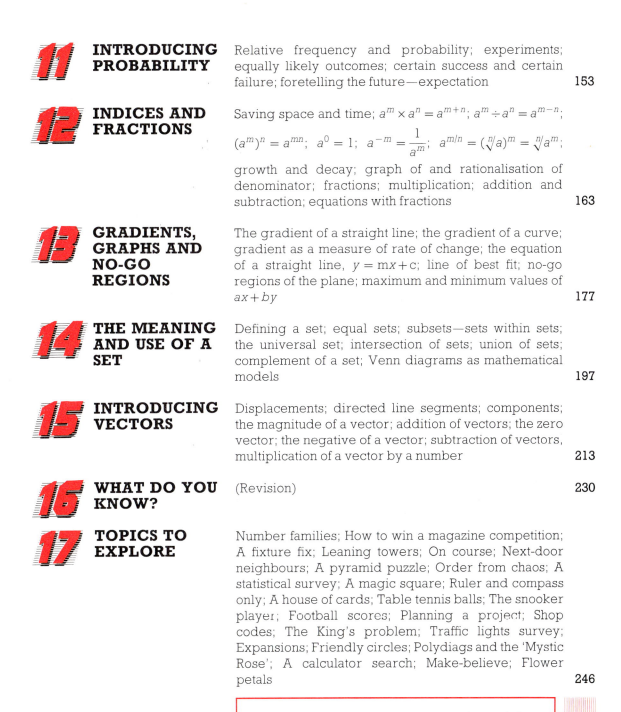

> **INVESTIGATIONS** will always be in a right-handed box

> **BRAINSTORMERS** will always be in a left-handed box

Mathematics is in action all around you. You need mathematics in your daily life. **Mathematics in Action** has been written to help you to understand and to use mathematics sensibly and well—to save you time and effort. Some parts of mathematics are needed in other subjects and some parts will be studied in greater detail later on.

Mathematics in Action follows the latest thinking in *what* mathematics should be studied and how hard, or easy, it should be. So you are taken forward, stage by stage, as far as you can go.

Exercises for practice, Puzzles and Games for fun, Brainstormers to make you think, Investigations to explore, Practical Activities, even Check-ups (to see how you're doing)—all are here.

Enjoy maths with **Mathematics in Action!** Let's hope that you will find a lot that is worthwhile, interesting, and above all, useful.

MiAG June 1988

Class discussion

The **car assembly line's function** is to take lots of parts and put them together to make a motor car.

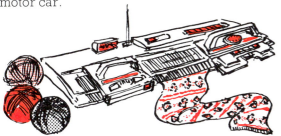

The **knitting machine's function** is to take separate balls of wool and use a pattern to turn out a cardigan or scarf.

The **chef's function** is to take the milk, flour, eggs and sugar, and to use the recipe to prepare a plate of pancakes.

Arithmetic: 10% of £15

Geometry: ⬚

Algebra: $5x - 6 > 3x$

Trigonometry: $\sin A = \frac{1}{2}$

This **textbook's function** is to help you to take all the ideas and techniques and put *Mathematics in Action*.

Each process can be summed up like this:

$$\text{Input} \rightarrow \boxed{\textbf{Function}} \rightarrow \textbf{Output}$$

Each input has a definite output. For example, the chef would expect his ingredients and recipe to produce pancakes, not apple pie!
Can you think of other examples of the *Input → Function → Output* idea?

MATHEMATICAL FUNCTIONS

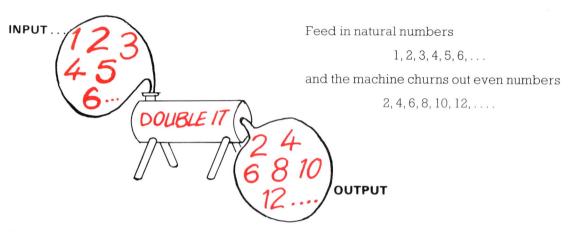

Feed in natural numbers

$$1, 2, 3, 4, 5, 6, \ldots$$

and the machine churns out even numbers

$$2, 4, 6, 8, 10, 12, \ldots$$

The function machine **maps** 1 to 2, 2 to 4, ... like this:

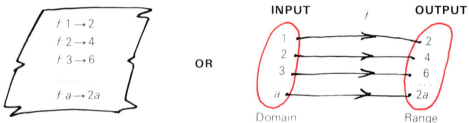

The numbers in the output are called the *values* of f. The value of f at 1 is 2, written $f(1) = 2$.
In the same way, $f(2) = 4$, $f(3) = 6$, ..., $f(a) = 2a$.
The **formula** for f is $f(x) = 2x$, where x is a natural number.

> A function from a set A to a set B is a rule which links each member of A with **one** member of B.

═══════════════ *Exercise 1* ═══════════════

1A List the output from each function machine.

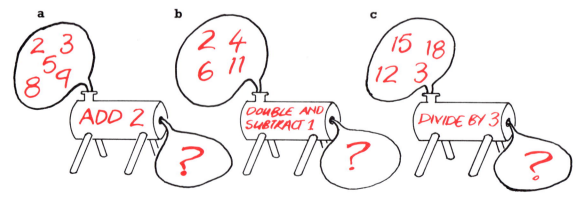

2A

Copy and complete:

Number of pence		Number of sweets
2 ●——f——→		● 4
4 ●————→		● 8
6 ●		●
8 ●		●
a ●		●

$f: a \rightarrow \dots$

3A Copy and complete:

Number of pence		Number of hours
50 ●——f——→		● 1
100 ●		●
150 ●		●
a ●		●

$f: a \rightarrow \dots$

4A The time, t seconds, to cover the measured mile at x mph is given by the formula $t(x) = \dfrac{3600}{x}$.

Calculate $t(60)$ and $t(180)$.

5A

For every bag of sand, the cement mixer makes three bags of concrete. The formula for its operation is $m(x) = 3x$.
Calculate $m(2)$, $m(4)$, $m(6)$.

6A To buy the stereo radio cassette recorder on HP, the formula for the cash paid in £s after x months is $c(x) = 25 + 1 \cdot 5x$.
Calculate $c(9)$, $c(12)$ and $c(15)$.

7A a Copy and complete a possible formula for a suitable function.
   b Calculate the value of each function at $x = 10$.

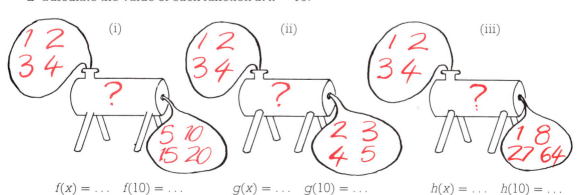

(i)   1 2 3 4 ? 5 10 15 20

(ii)   1 2 3 4 ? 2 3 4 5

(iii)   1 2 3 4 ? 1 8 27 64

$f(x) = \dots$ $f(10) = \dots$ $g(x) = \dots$ $g(10) = \dots$ $h(x) = \dots$ $h(10) = \dots$

8A

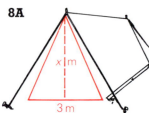

The formula for the area in square metres of the front of the tent is
$A(x) = \frac{3}{2}x$.
Calculate $A(1)$, $A(2)$ and $A(4)$.

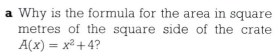

9B a Find a formula, in metres, for $P(x)$, the perimeter of the fence (without the gate).

b Calculate $P(12)$.

c Given $P(a) = 27$, find a.

10B

a Why is the formula for the area in square metres of the square side of the crate $A(x) = x^2 + 4$?

b Calculate $A(1)$, $A(2)$ and $A(3)$.

c Given $A(a) = 40$, find a.

Codes and code–breakers

A	B	C	D	E	F	G	H	I	J	K	L	M	N	O	P	Q	R	S	T	U	V	W	X	Y	Z
1	2	3	4	5	6	7	8	9	10	11	12	13	14	15	16	17	18	19	20	21	22	23	24	25	26

1 Zizi, the master spy has a coding formula
$c(x) = x + 5$.

a Check that the word CODE becomes
8|20|9|10.

b Check that the decoding formula is
$d(x) = x - 5$.

2 Another coding formula Zizi uses is $F(x) = x + 7$.
Write down the *decoding* formula for F. Use it to decode this conundrum:

30|15|8|27   25|16|21|14   16|26   26|24|28|8|25|12?
8   9|22|31|16|21|14   25|16|21|14.

3 A puzzle was coded using the formula $E(x) = 2x$. This code was then coded using the formula $F(x) = x + 7$.

a Find the decoding formula for this double code.

b Check it by double coding and then decoding.

c Test it on the following:

53|23|9|47   15|25|15   47|23|17   45|23|57   45|47|37|35|17   45|9|57?
25   53|25|45|23   25   53|9|45   9   31|25|47|47|31|17   11|37|31|15|17|43.

4 Make up some coding and decoding formulae of your own. Can you make them complicated enough to prevent a friend working them out?

GRAPHS

Exercise 2

1A Graphs come in all shapes and sizes, and every one tells a story. Often they can be used as mathematical models of events. Decide which of these graphs fits each story best.

a From the boundary Tom throws the cricket ball to the keeper.

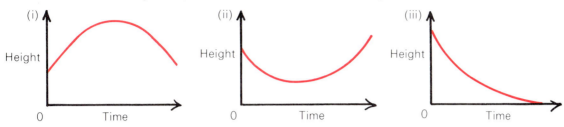

b The car accelerates away from rest, then stops at crossroads.

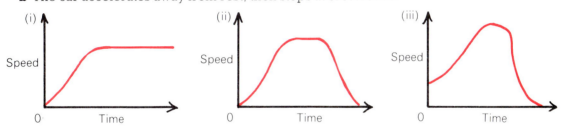

c A bean is planted. It germinates and grows. Its height is measured daily.

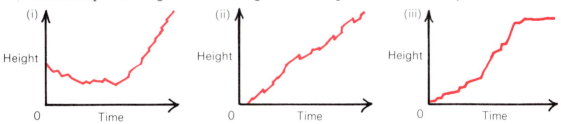

d The radiator is switched on in the room, and the temperature is measured hourly.

e Alice is blowing up a balloon.

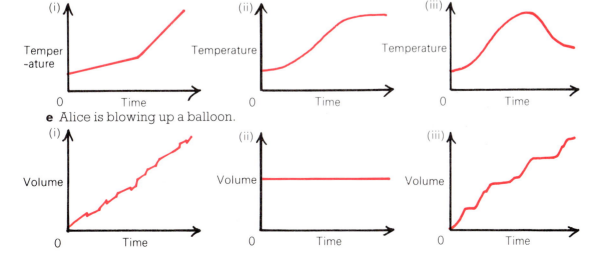

2A These two graphs show the heights of girls and boys up to age 18.

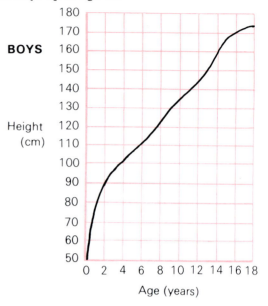

a Estimate the average heights of girls and boys at the age of:
   (i) 2   (ii) 4   (iii) 16
b Read off the average ages of girls and boys whose heights are:
   (i) 70 cm   (ii) 130 cm   (iii) 160 cm
c At what age do: (i) girls'   (ii) boys', heights begin to level off?
d Between what ages are both girls and boys growing fastest?
e Use the graph for girls' growth to copy and complete this table:

Age (years)	2	4	6	8	10	12	14	16	18
Height (cm)									

3A The graph shows the shortest stopping distances for cars on dry roads.

a Why does it start at (0,0)?
b Estimate the shortest stopping distances for speeds of 15 mph and 31 mph.
c Why does the curve rise from left to right?
d Estimate the speeds for stopping distances of 50 feet and 100 feet.
e Make a table of stopping distances for speeds of 10, 20, . . . , 50 mph.

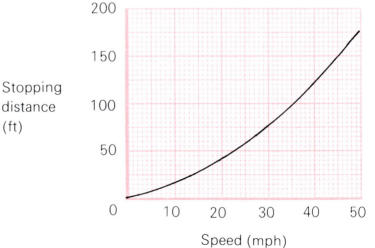

4B

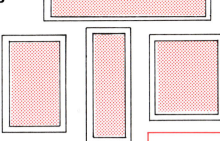

All of these windows let in the same amount of light. Each has an area of 6 m².
Which of the aluminium frames would be cheapest to make? Do you agree that this means looking for the one with least perimeter?
Draw a graphical model to find the answer, with the help of the questions below.

a Copy and complete the table.
($A = 6 = l \times b$.)

Length (m)	1	1·5	2	3	4	5	6
Breadth (m)	6	4			1·5		
Perimeter (m)	14		10			12·4	

b Draw a graph of perimeter against length, using the scales shown.
c What shape of window would cost least?

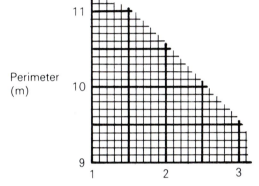

GRAPHS OF FUNCTIONS AS MATHEMATICAL MODELS

A graph gives a picture of a function. It shows the link between the sets of numbers in the *domain* (input) and the *range* (output).

The straight line is the graph of the function f defined by $f(x) = 2x$, for $0 \leqslant x \leqslant 5$.

$y = 2x$ **is the equation of the graph of the function f** $(0 \leqslant x \leqslant 5)$.

Graphs provide pictures, or mathematical models, of many practical situations.

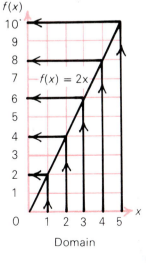

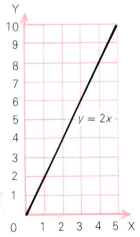

For example, a metal sheet is 5 m long and 2 m broad. If a part x m long is cut off, what is its area in m²?

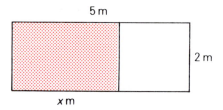

The area defines a function f where $f(x) = 2x \ (0 \leqslant x \leqslant 5)$. Associated with this practical situation of cutting off an area of metal, we have:

  (i) a *function f* given by $f(x) = 2x \ (0 \leqslant x \leqslant 5)$
 (ii) the *graph of the function* (shown on page 7)
(iii) the *equation of the graph*, $y = 2x \ (0 \leqslant x \leqslant 5)$.

Each of these can provide a **mathematical model** of the practical situation, as this Chapter shows.

I STRAIGHT LINES

$L: x \rightarrow ax + b$ Equation of graph, $y = ax + b$ Graph of L

  (i) In the equation $y = ax + b$, put $a = 1$, $b = 0$. The equation of the graph is $y = x$.

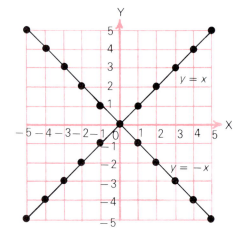

x	-5	-4	-3	-2	-1	0	1	2	3	4	5
y	-5	-4	-3	-2	-1	0	1	2	3	4	5

For $-5 \leqslant x \leqslant 5$, the graph is the straight line through the points given by the table of values.

 (ii) Putting $a = -1$, $b = 0$, the equation of the second graph is $y = -x$.

A function L defined by L: $x \rightarrow ax + b$ is called a linear function. Why do you think this is?

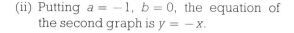

Exercise 3

1A Draw the graphs with equations: **a** $y = x$ **b** $y = -x$, for $-6 \leqslant x \leqslant 6$.

2A On the same sheet as question **1A**, draw the graphs with equations:
    a $y = 2x$ **b** $y = 3x$ **c** $y = -2x$ **d** $y = -3x$.

3A On the same sheet again, draw the graphs with equations:
    a $y = 4x$ **b** $y = \frac{1}{2}x$ **c** $y = -6x$.

4A Write a sentence to describe the graphs in questions **1A–3A**.

5A Since graphs with equations $y = ax+b$ are straight lines, to draw them we only need to plot two points. To draw the graph of $y = x+2$, find the points where the line crosses the x- and y-axes.

When $x = 0$, $y = 0+2 = 2$, giving the point $(0, 2)$.
When $y = 0$, $0 = x+2$, so $x = -2$, giving the point $(-2, 0)$.
Join the points $(0, 2)$ and $(-2, 0)$.

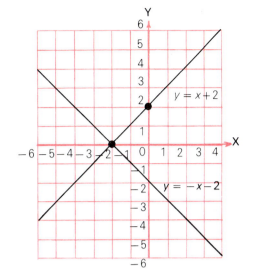

a Draw the graphs of $y = x+2$ and $y = -x-2$, from $x = -6$ to $x = 4$.

b On the same sheet, draw the graphs of:
  (i) $y = x+1$ (ii) $y = -x-1$
  (iii) $y = x+3$ (iv) $y = -x-3$

c Write a sentence or two to describe the graphs with equations $y = x+1$, $y = x+2$ and $y = x+3$.

6B $L: x \rightarrow ax+b$. Make a rough sketch of the graph of L for:
  (i) $a = 1, b = 0$ (ii) $a > 0, b = 0$ (iii) $a < 0, b = 0$
  (iv) $a = 1, b > 0$ (v) $a = 1, b < 0$.

Using graphs of linear functions (straight lines) as mathematical models

=================== *Exercise 4* ===================

1A Hiram III 'had done' Europe, and was off home. First he had to change his British pounds to American dollars.

a Copy and complete this exchange table:

Number of £s, p	0	10	20	30	40	50
Number of $, d	0	15	30			

b Draw a graph on $\frac{1}{2}$-cm squared paper, taking 2 squares to 10 units on each axis, and number of £s on the horizontal axis.

c Read off the number of dollars for £45.

d Copy and complete the equation of the graph $d = \ldots p$.

2A

Dan is practising his diving. His diving speed is measured every second.

a Draw a graph of s against t for $0 \leqslant t \leqslant 2$.

b Read off Dan's speed after:
   (i) 0·25 seconds
   (ii) 1·75 seconds.

c Copy and complete:
   $s = \ldots t$.

Time, t seconds, from the start	Speed s m/s
0	0
0·5	5
1	10
1·5	15
2	20

3A Match each graph with one of the water tanks. Explain your choice.

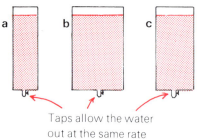

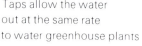

Taps allow the water out at the same rate to water greenhouse plants

(Time after the taps have been opened)

4B When moist air rises its temperature falls 5°C every kilometre. At sea level the temperature is 10°C.

a Copy and complete:

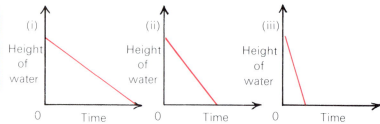

Height, H km	0	1	2	3	4
Temperature, T°C	10	5			

b Draw the graph of T against H, for $0 \leqslant H \leqslant 4$.
c What is the temperature at a height of: (i) 1·5 km   (ii) 5 km?

5B Which graph is most likely to 'model' the situation as the kitchen paper is unrolled? Give reasons for your answer.

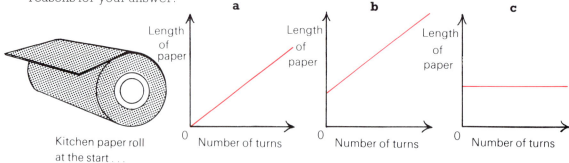

Kitchen paper roll at the start . . .

When Hiram III first arrived in England he wanted to hire a car for a day. He was puzzled about the best choice from these three firms. Being an Associate Professor of Mathematics in the USA he decided to solve his problem with a mathematical model which would calculate his total cost £y per day for motoring x miles.

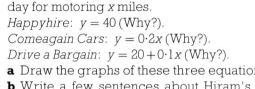

Happyhire: $y = 40$ (Why?).
Comeagain Cars: $y = 0.2x$ (Why?).
Drive a Bargain: $y = 20 + 0.1x$ (Why?).

a Draw the graphs of these three equations on the same sheet, for $0 \leqslant x \leqslant 300$.

b Write a few sentences about Hiram's model, and what it told him. How would you advise him about which firm he should choose?

II PARABOLAS

$Q: x \rightarrow ax^2 + bx + c$ ⟩ Equation of graph, $y = ax^2 + bx + c$ ⟩ Graph of Q

In the equation $y = ax^2 + bx + c$, put $a = 1$, $b = c = 0$.
The equation of the graph is $y = x^2$.

x	-5	-4	-3	-2	-1	0	1	2	3	4	5
y	25	16	9	4	1	0	1	4	9	16	25

The graph is a *parabola* through the points given by the table of values. The minimum value of Q is $Q(0) = 0$.

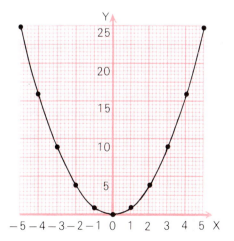

A function Q defined by $Q(x) = ax^2 + bx + c$ is called a quadratic function.

=== *Exercise 5* ===

1A a Draw the graph with equation $y = x^2$, using the table of values, axes and scales above.

 b The parabola has an axis of symmetry. Describe it.

2A a Draw the graph of the function Q with equation $y = x^2 + x$ for $-4 \leqslant x \leqslant 4$. First copy and complete a table of values, row by row.

x	-4	-3	-2	-1	0	1	2	3	4
x^2	16								
$y = x^2 + x$	12								

or

x	-4	-3	-2	-1	0	1	2	3	4
$y = (x+1)x$	12								

Using a calculator, when $x = -4$,

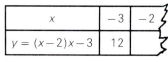

$(x+1)x = \boxed{-4} + \boxed{1} = \times \boxed{-4} = \boxed{\qquad 12}$

b Draw the line of symmetry.

c Write down the minimum value of Q, and the corresponding value of x.

3A a Draw the graphs of the functions Q_1 with equation $y = x^2 - 4$ and Q_2 with equation $y = 4 - x^2$, for $-3 \leqslant x \leqslant 3$.

b Describe the axis of symmetry.

c Write down the maximum and minimum values of Q_1 and Q_2, and the corresponding values of x.

4A a Draw the graph of the function Q with equation $y = x^2 - 2x - 3$, for $-3 \leqslant x \leqslant 5$. Use one of these tables of values:

b Draw the axis of symmetry.

c Write down the minimum value of Q, and the corresponding value of x.

x	-3	-2
x^2	9	
$-2x$	6	
-3	-3	
$y = x^2 - 2x - 3$	12	

OR

x	-3	-2
$y = (x-2)x - 3$	12	

(Using your calculator)

5B a Draw the graph of the function Q with equation $y = 6 - x - x^2$, for $-4 \leqslant x \leqslant 4$. If you use a calculator, take $y = (-1 - x)x + 6$.

b Draw the axis of symmetry.

c Write down the maximum value of Q, and the corresponding value of x.

6B $Q: x \rightarrow ax^2 + bx + c$. Investigate the shape of the graph of Q (a parabola) for:

(i) $a > 0, b = c = 0$ (ii) $a < 0, b = c = 0$ (iii) $a > 0, b = 1, c = 0$

(iv) $a > 0, b = -1, c = 0$ (v) $a > 0, b = 2, c = 0$ (vi) $a < 0, b = -2, c = 0$.

Using graphs of quadratic functions (parabolas) as mathematical models

Exercise 6

1A

The life–raft falls distance d metres in t seconds, as shown in the table:

t	0	1	2	3	4
d	0	5	20	45	80

a Draw the graph of d against t, from $t = 0$ to $t = 4$ using the scales shown.

MATHEMATICAL MODELS—FUNCTIONS AND GRAPHS

b The graph looks like a parabola through O, so its equation will be $d = kt^2$. Check this from the table of values, and find k.

c Now calculate: (i) d when $t = 6$
(ii) t when $d = 100$.

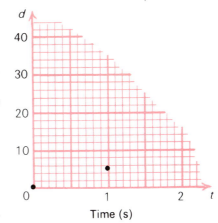

Distance (m)

Time (s)

2A

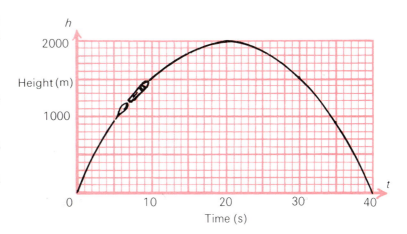

An architect is designing a bridge which is supported by a parabolic arch.

As a mathematical model, she chooses the equation

$y = 6x - x^2$ [$= (6 - x)x$ for calculator].

a Make a table of values for $x = 0, 1, 2, ..., 6$.
b Draw the graph of the parabola.
c The unit is the metre. From your graph estimate the height and width of the arch.

3B The graph shows the parabolic flight–path of a rocket on trial.
  a Read off:
    (i) the maximum height, and the time to this height
    (ii) the total flight time
    (iii) the height after 35 seconds
    (iv) the times for a height of 1500 m.

Height (m)

Time (s)

b The engineers estimate that the equation of the parabola is $h = kt - 5t^2$: h is the height in metres after t seconds. Find k.

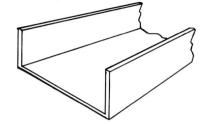

Jim's job is to take a sheet of metal 20 cm wide and bend the sides up to make a channel for carrying water. The ends have to be rectangular. 'Make the channel as big as possible', he's told. How can he do this? Trial and error? A mathematical model?

a Trial and error

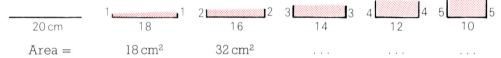

	20 cm	18	16	14	12	10
Area =		18 cm²	32 cm²	. . .	. . .	. . .

Copy his sketches, and draw more of them. Investigate the maximum area, and the height and width of the channel then.

b A mathematical model

$A = x(20 - 2x)$. Why?

Draw the graph of A against x, and find the dimensions of the channel for greatest cross–sectional area.

c Investigate the dimensions of a V-shaped channel made from the same sheet of metal which give the maximum cross–sectional area.

Practical

1 Construct the 'envelope' of a parabola by drawing two lines of equal length at any angle, dividing them up equally, and joining pairs of points as shown. Try different angles, and more divisions on the lines. 'Curve stitching', using wool stitched through card, produces striking designs.

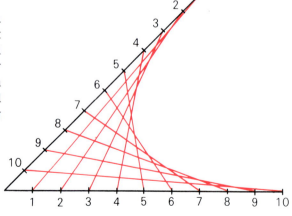

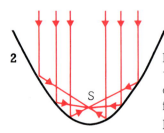

2 Investigate how a telescope mirror gathers rays of starlight. Use your graph of the parabola $y = x^2$ (Question **1A** of Exercise 5) to draw the parallel rays and their reflections, which meet at the focus S. Investigate also the way in which torches and car headlights project cylindrical beams of light.

III $\boxed{C: x \rightarrow ax^3 + bx^2 + cx + d}$ ⟩ $\boxed{\text{Equation of graph, } y = ax^3 + bx^2 + cx + d}$ ⟩ $\boxed{\text{Graph of } C}$

In the equation $y = ax^3 + bx^2 + cx + d$, put $a = 1, b = c = d = 0$.
The equation of the graph is $y = x^3$.

x	-3	-2	-1	0	1	2	3
y	-27	-8	-1	0	1	8	27

A function C defined by

$$C(x) = ax^3 + bx^2 + cx + d$$

is called a *cubic function*. Why cubic?

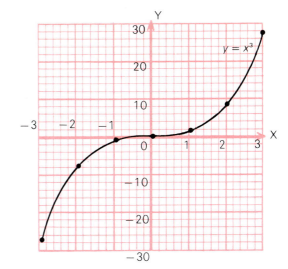

=== *Exercise 7* ===

1A a Draw the graph with equation $y = x^3$, using the table of values, axes and scales above.
   b The graph is symmetrical about O. Check this (tracing paper might help).

2A a Draw the graph of the function C with equation $y = x^3 - 4x$ for $-3 \leqslant x \leqslant 3$. But first copy and complete:

x	-3	-2
x^3	-27	
$-4x$	12	
$y = x^3 - 4x$	-15	

OR

x	-3
$y = (x^2 - 4)x$	-15

(Using your calculator)

   b Read off:
      (i) the maximum and minimum turning values of C, and the corresponding values of x
      (ii) the values of x for which the value of C is zero.

3A a Draw the graph of the function C with equation $y = x^3 - 3x^2$, for $-2 \leqslant x \leqslant 4$.
   b Read off:
      (i) the maximum and minimum turning values of C, and the corresponding values of x
      (ii) the values of x for which the value of C is zero.

4B a Draw the graph of the function C with equation $y = x^3 - x^2 - 2x$, for $-3 \leqslant x \leqslant 4$. Using a calculator, $y = [(x-1)x - 2]x$.
   b Answer the same questions as in **3A b** (i) and (ii).

IV HYPERBOLAS

$H: x \rightarrow \dfrac{a}{x}$ ⟩ Equation of graph, $y = \dfrac{a}{x}$ ⟩ Graph of H

In the equation $y = \dfrac{a}{x}$, put $a = 24$. The equation of the graph is $y = \dfrac{24}{x}$.

x	-24	-16	-12	-8	-6	-4	-3	-2	$-1{\cdot}5$	-1	0	1	$1{\cdot}5$	2	3	4	6	8	12	16	24
y	-1	$-1{\cdot}5$	-2	-3	-4	-6	-8	-12	-16	-24	$-$	24	16	12	8	6	4	3	2	$1{\cdot}5$	1

The graph is called a *hyperbola*. You'll see this curve on the wall of a room lit by a standard lamp.

The further you go along the axes in either direction the closer the curve approaches them. The axes are called *asymptotes* to the curve.

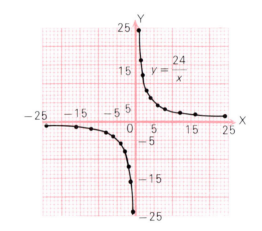

$y = \dfrac{24}{x}$

=== *Exercise 8* ===

1A a Draw the graph with equation $y = \dfrac{24}{x}$, using the table of values, axes and scales above.

  b Mark: (i) the axes of symmetry (ii) the centre of symmetry.

  c Calculate y when $x = 100, 1000, 1\,000\,000$. Copy and complete:
    'As x increases, $y \ldots\ldots\ldots$, and the graph gets closer and closer to $\ldots\ldots\ldots$'

  d Calculate y when $x = 0{\cdot}01, 0{\cdot}001, 0{\cdot}000\,001$. Copy and complete:
    'As x decreases, $y \ldots\ldots\ldots$, and the graph $\ldots\ldots\ldots$'

2A a Draw the graph of the hyperbola $y = \dfrac{20}{x}$, by making a table of values with $x = -20, -10, -5, -4, -2, -1, 1, \ldots, 20$.

  b Describe as many special features of the graph as you can.

3B a This time the equation of the hyperbola is disguised as $xy = -20$. Draw the graph, using the same values of x as in question **2A**.

  b Describe the graph, and how it differs from previous ones.

4B Sketch hyperbolas with equations $y = \dfrac{a}{x}$ for: (i) $a > 0$ (ii) $a < 0$.

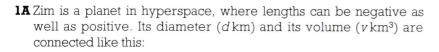

1A Zim is a planet in hyperspace, where lengths can be negative as well as positive. Its diameter (d km) and its volume (v km³) are connected like this:

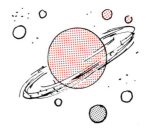

d	-4	-3	-2	-1	0	1	2	3	4
v	-32	-13.5	-4	-0.5	0	0.5	4	13.5	32

a Draw the graph of v against d on 2 mm squared paper, taking scales of 1 cm to 1 unit horizontally for d and 1 cm to 5 units vertically for v.

b (i) Look through the graphs that you have already drawn. Which one matches this one most closely?

(ii) Check from the table of values that the equation of the graph is $v = \frac{1}{2}d^3$.

(iii) Calculate v when $d = 10$.

2A

To check its petrol consumption, the Ranger Group's new model XX was driven round the test track at different speeds.

Speed, s m/s	5	10	20	30	40	50	60
Time, t seconds	240	120	60	40	30	24	20

a Draw the graph of t against s, for $5 \leqslant s \leqslant 60$. ($s$ on the horizontal axis).

b From your graph, estimate the time for a speed: (i) 15 m/s (ii) 25 m/s.

c (i) Look over the graphs you've drawn elsewhere. Which match this one most closely?

(ii) Using the table, estimate the equation of your graph.

3B Marzipan Metal Co. make *open* water tanks. Their design is based on square sheets of metal with squares of side x metres cut from the corners.

a Prove that the formula for the volume, V m³, of a tank is $V = 4x^3 - 40x^2 + 100x$.

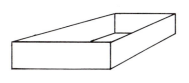

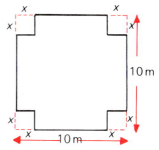

The Company realises that there are many possible shapes of box—short and fat, tall and skinny. How can they find the one with the greatest volume? They turn to mathematics.

b Graph V against x, for $x = 1$, $1\frac{1}{2}$, 2, $2\frac{1}{2}$ and 3.

c From the graph read off the greatest value of V and the corresponding value of x. Now write down the dimensions of the tank they should make.

Investigate the shape, symmetry and any other special features of the graphs with equations:

a $y = \sqrt{x}$ and $y = -\sqrt{x}$ on the same diagram

b $y = 2^x$ **c** $y = \dfrac{36}{x^2}$

d $y = \sqrt{(16-x^2)}$ and $y = -\sqrt{(16-x^2)}$ on the same diagram.

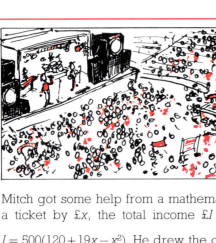

Mitch organises rock concerts. He knows that if he sells tickets at £5 each he will have a full house—12 000 people. He estimates that each £1 extra on a ticket will reduce the attendance by 500. What should he charge to make most money? Can you help him? You may be surprised by what you find.

Mitch got some help from a mathematical friend, and found: If he raised the price of a ticket by £x, the total income £I (number attending × cost per ticket) would be

$I = 500(120 + 19x - x^2)$. He drew the graph of $\dfrac{I}{500}$ against x, for $x = 0, 2, 4, 6, \ldots, 14$,

and found the greatest income and the value of x.

a Prove that $I = 500(120 + 19x - x^2)$.

b Draw the graph for $0 \leqslant x \leqslant 14$. (Why $\dfrac{I}{500}$?)

c Find the maximum income, and the corresponding ticket price.

CHECK-UP ON **FUNCTIONS AND GRAPHS**

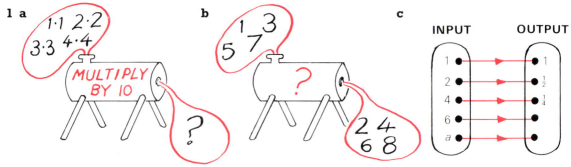

1 a List the output.

b Copy and complete a possible formula $f(x) = \ldots$

c Copy and complete the diagram.

2 Match each graph with its equation:

a $y = x$
b $y = -2x + 1$
c $y = 3x + 2$
d $y = -x$

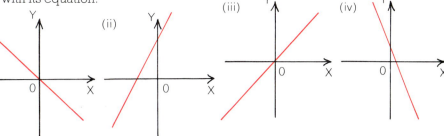

(i)   (ii)   (iii)   (iv)

3

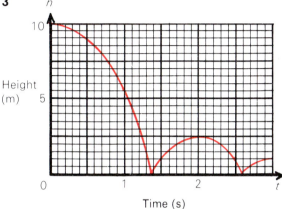

Time (s)

A rubber ball is dropped onto the floor.
a When did the ball first hit the surface?
b What height was its first bounce?
c Describe what happens as time increases.
d Using the graph, make a table of values of height for $t = 0, 0.5, 1.0, 1.5, 2.0, 2.5$.

4 Match each graph with one of the equations.

a $y = ax^2 + bx + c$
b $y = \dfrac{a}{x}$
c $y = ax + b$
d $y = ax^3 + bx^2 + cx + d$

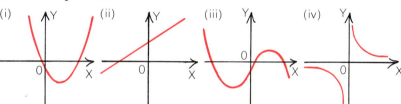

(i)   (ii)   (iii)   (iv)

5 Caring Car Company make twenty cars each week: x cars are 2-door, and y cars are 4-door.
a Write down a formula for y in terms of x.
b Draw a graph of the formula.

6

Hannah and Hal rent 100 m² of floor area in the Indoor Craft Market. Which rectangular shape should they choose to keep the wall space to a minimum? Make a mathematical model. Take x metres for the length.

a Prove that the floor's perimeter, P m, is $P = 2\left(x + \dfrac{100}{x}\right)$.
b Draw a graph, taking $x = 4, 6, 8, 10, 12, 14, 16$.
c Find x for the minimum perimeter (and wall space).

SUPERANNUATION, NATIONAL INSURANCE AND INCOME TAX

Meet Mr Dixon, a car mechanic to trade. As you can see, he's hard at work.
Here is his pay-slip for a recent week.

EMPLOYEE 123	**J. DIXON**	TAX CODE 379H	WEEK 52
BASIC PAY 180·00	OVERTIME 20·00	COMMISSION —	GROSS PAY 200·00
INCOME TAX 31·60	SUPERANNUATION 10·00	NATIONAL INSURANCE 18·00	TOTAL DEDUCTIONS 59·60
			NET PAY

Exercise 1

1A a What items make up: (i) his Gross Pay   (ii) the deductions from his pay?
  b How much is his Net Pay, or 'take-home' pay?
2A Study the drawing below, and then check the entry on the pay-slip for superannuation.
Mr Dixon pays 5% of his Gross Pay for superannuation.

3A Mrs Dixon works in a department store, and earns £660 a month. She pays 6% of this for
superannuation. How much does she pay: **a** per month   **b** in a year?
4A The Dixons' daughter Ann works for a computer firm, and has just had her monthly salary
increased to £1138. She now pays £62·59 superannuation monthly. What percentage is this
of her monthly pay?

5A Employers also make contributions to their employees' superannuation schemes. Mr Dixon's employer pays an amount equal to 7% of Mr Dixon's pay. Including Mr Dixon's contributions (5% of his pay), calculate the total sum paid for his superannuation in: **a** 1 year **b** 30 years.

Look at Mr Dixon's pay-slip again. You'll see a deduction for National Insurance (NI).

what is 'National Insurance' for?

who pays it?

How much do I pay?

Sickness Benefit unemployment Benefit State Pensions Maternity Grant Death Grant Redundancy Fund

All employees who earn more than a certain amount. Employers also contribute

GROSS INCOME		
Weekly	Monthly	Rate %
£0–£37·99	Under £165	0
£38–£59·99	£165–£259·99	5
£60–£94·99	£260–£411·99	7
£95–£285	£412–£1235	9
over £285	over £1235	9% of £285, or of £1235

6A Check the deduction of £18 on Mr Dixon's pay-slip.

7A How much NI would Mr Dixon pay for
Gross Pay of: **a** £35 in week 1 **b** £90 in week 2
c £142·30 in week 3?

8A Calculate Ann's NI contributions in months
when her gross salary was:
a £820 **b** £1235 **c** £1320

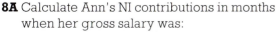

9A Teachers, and employees in the Health Service and in most large companies, pay different rates from the ones in the table. Mr Jones earns £100 a week. His NI contribution is 9% of the first £38 he earns, and 6·85% of the remainder.

    a Calculate his weekly contribution.

    b How does this compare with someone in the Government scheme who earns £100 a week?

10B Paul Power earned £37·50 in a part-time job in week 45. The following week he worked longer hours, and earned £39. He was rather annoyed when he received week 46's pay-slip. Why?

Now for Income Tax. Read through this conversation, and then study the example on the next page to see how it works out.

EMPLOYEE 123	**J. DIXON**	TAX CODE 379H	WEEK 52
BASIC PAY 180·00	OVERTIME 20·00	COMMISSION —	GROSS PAY 200·00
INCOME TAX 31·60	SUPERANNUATION 10·00	NATIONAL INSURANCE 18·00	TOTAL DEDUCTIONS 59·60
			NET PAY

What is 'Income Tax' for?

To pay for Education, Housing, Defence, Health, Law and Order, Social Security, etc.

Who pays tax?

Everyone who earns more than a certain amount

What income is taxed?

Earnings, investments, pensions, etc

Is the whole of my income taxed?

NO. Tax free allowances are: Single person's: £2425 (1987-88) Married man's: £3795 Wife's earned income: £2425 .... and others ...

What is taxable income?

Total income minus allowances

How is income tax paid?

Your employer deducts it from your pay and passes it to the Tax office. This is called Pay As You Earn (P.A.Y.E)

How much tax have I to pay?

On a taxable income of £1 to £17,900, 27p in the £ (27%). Higher rates for bigger incomes

What is a tax code?

The Tax office sends it to your employer to say how much you have to pay. Code 379 means you have about £3790 - 3799 tax free allowance

Example

Check the income tax deduction in Mr Dixon's pay-slip. Assume that his Gross Pay is £200 every week, and that tax is calculated on his annual pay, less superannuation and allowances.

Annual Pay = 52 × £200 = £10 400
Superannuation = 5% of pay = £ 520
_____
£ 9880
Married Man's allowance = £ 3795
_____
Taxable income = £ 6085
_____

Tax to pay = 27% of £6085 = £1642·95 for the year.
Weekly tax deduction = £1642·95 ÷ 52 = £31·60.

═══════════ *Exercise 2* ═══════════

1A

Rob is an engineer. He is single, and earns £190 a week. Copy and complete:

Annual pay = 52 × £190 = £ . . .
Single person's allowance = . . .
_____
Taxable income = . . .
_____
Income tax due = 27% of £ . . . = . . . per year
= . . . per week.

2A Shona Bates is a waitress. In her annual income tax return she enters her wage as £3540, and 'tips' (also taxable) as £230. Calculate the amount of tax she has to pay:
a annually **b** weekly.

3A

Fred earns £250 gross, weekly, as a garage foreman. Calculate:
a his Gross Pay for a year
b his taxable income (after a single person's allowance)
c the tax he pays: (i) annually (ii) weekly
d his net weekly pay.

4A Calculate the amount of income tax paid annually by:
a Mr T. Jones, married, annual salary £15 105
b Miss M. Taylor, annual salary £11 980
c Mr B. Davidson, single, weekly pay £163·10
d Mrs A. Burns, weekly pay (part–time job) £49·50.

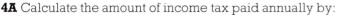

TAKE-HOME PAY

5A Copy and complete all the empty boxes in Richard Richman's monthly salary slip. Assume that his basic pay, commission and National Insurance are the same every month, and that he has an allowance of ten times his tax code number. What is his take–home pay for the month?

EMPLOYEE 00001	**RICHARD RICHMAN**		TAX CODE 365H		MONTH 4	
BASIC PAY 1325·25	OVERTIME —	COMMISSION 170·75		BONUS —		GROSS PAY ?
INCOME TAX ?	SUPERANNUATION —	NATIONAL INSURANCE 111·15		OTHER DEDUCTIONS —		TOTAL DEDUCTIONS ?
	NET PAY ?					

6A Robin and Katie Laidlaw are a married couple. Robin earns £15 350 per annum as an accountant. Katie earns £4850 per annum as a part-time nurse. She has the 'wife's earned income' allowance. Calculate the total tax deducted from their salaries.

7B Mr Dixon's employer sends him an *annual* statement, showing the total deductions for superannuation, National Insurance and income tax. Copy and complete this statement, based on information given on the last few pages. Tax is calculated on Gross Pay less superannuation and allowances.

Name J. DIXON		Tax code 379H
TOTALS FOR YEAR		
Gross Pay	Superannuation	Gross Pay less Superannuation
Tax Allowance	Taxable Income	Tax deducted
Employee's NI Contribution	Tax + NI deducted	Net Pay

Find out: **a** when the financial year starts and ends

   b who gives the 'Budget' speech

   c when and where the speech is made

   d what the present allowances and rates of income tax are. Why do these change nearly every year?

A higher rate of tax for big earners (1988–89)

In the 1988 Budget the Chancellor of the Exchequer changed the various allowances, and simplified the system of higher rate taxes.

For 1988–89 the allowances were:

Single person's allowance	£2605
Married man's allowance	£4095
Wife's earned income allowance	£2605

The rates of tax became:

Taxable income (£)	Rate %	
1 to 19 300	25	. . . the basic rate of tax
over 19 300	40	. . . the higher rate of tax.

Example

Action Rovers' top goal scorer Mac James earns £25 000 a year.
Calculate his monthly tax deductions under PAYE.

Annual income   = £25 000
Single person's allowance =   2605
$$\overline{\text{Taxable income}} = £22\,395$$

Tax to pay: (i) £19 300 at 25% = £4825
         (ii) £22 395–£19 300
              = £3095 at 40% = £1238
                  TOTAL = £6063

Monthly tax deduction = £6063 ÷ 12 = £505·25.

Exercise 3

Use the 1988–89 allowances and rates of tax in this Exercise, unless others are given.

1A Anne Scott's new job gave her a salary of £24 000 a year. Calculate her monthly tax bill.

2A Mr Keene, single, was promoted, and his gross weekly pay became £475. Calculate the tax deducted from his weekly pay packet. (You'll have to calculate the tax on his gross annual pay first.)

3A During one tax year Sam Swinger's earnings as a professional golfer were £126 540. His tax allowances totalled £18 500.
   a How much income tax had he to pay?
   b How could he have such a high tax allowance?

4A Copy and complete this table to find out what percentage of their weekly pay single people with different incomes pay in tax.

Gross weekly pay (£)	Gross annual pay (£)	Taxable income (£)	Tax payable per year (£)	Tax payable per week (£)	Tax as a percentage of pay
50	2600	0	0	0	0
100					
200					
600					

5B

This happy couple have done their tax sums—their wedding day is 6th April, the start of the new tax year. In his first year of married life how much *less* tax will the bridegroom pay on a salary of:

a £6000   **b** £14 300   **c** £25 000,

than if he had been single?

6B Jim came across this flow-chart which was used before the changes in tax rates were made in 1988. Use it to investigate the taxable income bands and rates of tax.

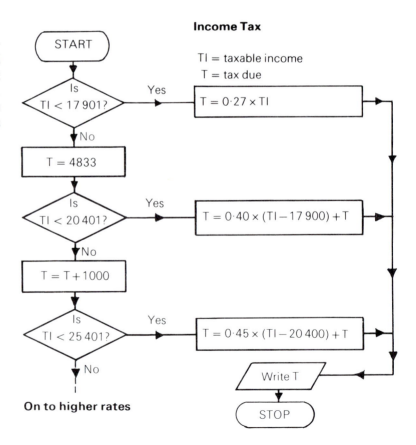

Income Tax

TI = taxable income
T = tax due

START

Is TI < 17 901? — Yes → T = 0·27 × TI

No

T = 4833

Is TI < 20 401? — Yes → T = 0·40 × (TI − 17 900) + T

No

T = T + 1000

Is TI < 25 401? — Yes → T = 0·45 × (TI − 20 400) + T

No

On to higher rates

Write T

STOP

National income and expenditure, 1986–87

=========================== *Exercise 4* ===========================

1A Total income £155·9 billion (thousand million).

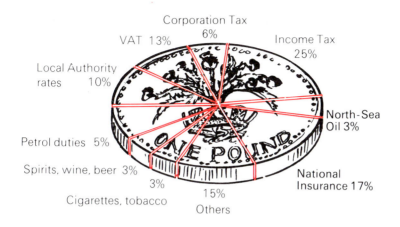

a Write out the total income in figures, in full. (You'll need a lot of zeros!)
b Make a list of all the items, in order, largest first.
c How much did: (i) income tax (ii) North Sea oil, raise?
d Do any of the figures surprise you?

2A Total expenditure £163·4 billion.

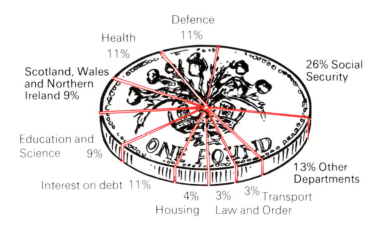

a Make a list of the items, in order, largest first.
b How much was spent on Education and Science?
c Do any of the figures surprise you?
d Did income tax raise enough money to cover Social Security spending?

3B Calculate the difference between the total national income and expenditure. How does the Government cope with this?

INCOME SUPPORT

Many people with low incomes receive weekly payments from the Government, called Income Support. In 1986 the payments were:

For a person living alone: £29·80.

For a couple with no children: £48·40.

For households with children living at home, extra money is paid; the amount depends on their ages.

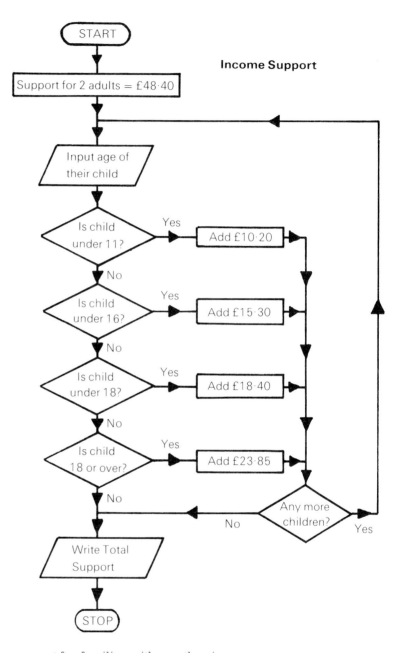

The flowchart shows the weekly support for families with no other income.

1A Each of the families in this house is due Income Support.
- **a** Which family lives in which flat?
- **b** How much should each family receive?
  - (i) Ian Thomson lives alone.
  - (ii) The Whitleys have no children.
  - (iii) Mr and Mrs Roberts have a 14-year-old son.
  - (iv) Mr and Mrs Jones have three children aged 4, 5 and 7.
  - (v) The Parry's children, still at school, are aged 15, 17 and 18.

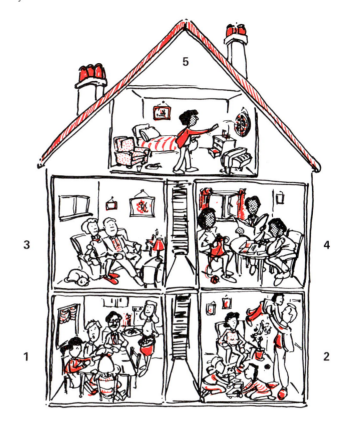

2A Mr and Mrs Brook receive £68·80 weekly. How many children have they, and how old are the children?

3B Mr and Mrs Summer receive £79 weekly. How many children are in the family? Is there only one possible answer? Explain.

Find out about the support and free services available to people on low incomes. Comment on the way in which society helps them.

1 Gavin James earns £148 a week. He pays 6% superannuation and 5% National Insurance.
  a Calculate how much he pays for each weekly.
     He also pays £27·50 tax and £3·25 union dues each week.
  b What is his weekly take–home pay?

2 Tahir, a married man, earns £14 275 a year. He pays income tax at the basic rate of 27p in the £. His personal allowances total £3840. Calculate:
  a how much income tax he has to pay
  b how much more tax he will pay after a 5% rise in his salary.

3

Employee 01002	Robert Keene		Tax code 365H	Month 7
Basic pay 1685·25	Overtime —	Commission 184·75	Bonus —	Gross Pay
Income tax	Superannuation —	National Insurance 117·50	Other deductions —	Total deductions
				Net pay

Mr Keene has £3655 allowances. He is taxed at 27p in the £ on £17 900 taxable income, and at 40p in the £ on the rest. His salary, commission and NI remain the same each month. Copy and complete his salary slip.

4 Mr and Mrs Williams have Income Support. They receive £48·40, plus £15·30 for each of their three children. How much do they receive altogether? What are the possible ages of the children?

5 Joan is in a job-sharing scheme in a city store. Calculate the entries **a–e** in her *annual* tax statement. What was her take–home pay for the year?

Name J. MILNE		Tax Code 233L
TOTALS FOR YEAR	Superannuation	Gross Pay minus Superannuation **a**
Gross Pay £4920	£196·80	Tax deducted **c**
Tax Allowance £2335	Taxable Income **b**	Net Pay **e**
Employee's NI Contributions £344·40	Tax and NI deducted **d**	

Class discussion—scale models and scale factors

1 What happens when you drop a stone into a calm pool of water?
The ripples move out as concentric circles.
Circles B, C and D are *enlargements* of A.
Circles A, B and C are *reductions* of D.

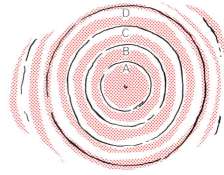

2

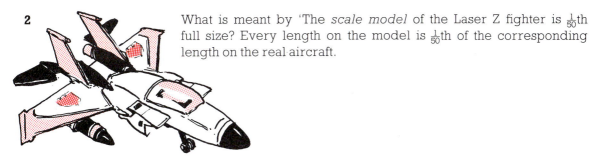

What is meant by 'The *scale model* of the Laser Z fighter is $\frac{1}{50}$th full size? Every length on the model is $\frac{1}{50}$th of the corresponding length on the real aircraft.

The model is a *reduction* in the ratio $1:50$—the *scale factor* from real aircraft to model is $\frac{1}{50}$.
Going the other way, the real aircraft is an *enlargement* in the ratio $50:1$—the *scale factor* from model to real aircraft is $\frac{50}{1}$.

3 In each diagram you see the real-life picture and its scale picture.

Reduction of the high–jumper

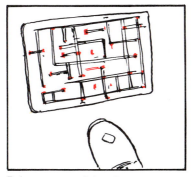

Enlargement of the silicon chip

Reduction of the sailing ship

When an object is reduced or enlarged, all the corresponding lengths are in the same ratio—they have the same scale factor.
In an enlargement, the scale-factor is greater than 1.
In a reduction, the scale factor is less than 1.

4 Andrew's picture on the screen looks real.
This is because in an enlargement (or in a reduction):

(i) corresponding angles are equal, and

(ii) corresponding lengths are in the same ratio, that is have the same scale factor.

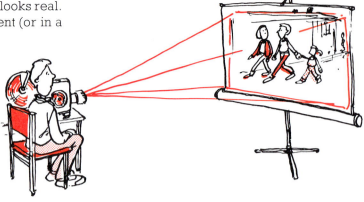

Example

The actual length of this car is 4·5 metres. The corresponding length in a building kit model is 30 cm. What is the scale factor for the kit?

$$\text{Scale factor} = \frac{\text{Kit length}}{\text{Actual length}} = \frac{30\,\text{cm}}{4\cdot5\,\text{m}} = \frac{30\,\text{cm}}{450\,\text{cm}} = \frac{1}{15}.$$

Practical

a Measure the length of: (i) a line on a photographic slide, or an overhead projector slide

                      (ii) the enlargement of the line on a screen, or wall.

b Calculate the scale factor of the enlargement.

c Try this for different lines. What do you find?

Exercise 1

1

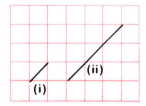

Calculate the scale factor of:

a the enlargement from (i) to (ii)

b the reduction from (ii) to (i).

2 Calculate the scale factor of:

  a the enlargements from:

    (i) to (ii), (ii) to (iii) and (i) to (iii)

  b the reductions from:

    (ii) to (i), (iii) to (ii) and (iii) to (i).

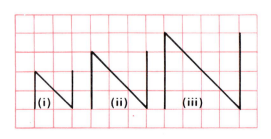

3 On squared paper:

  a *enlarge* the lengths of the sides of shape A by scale factor 3, B by scale factor 2, C by scale factor 1 and D by scale factor 4

  b *reduce* the lengths of the sides of shape E by scale factor $\frac{1}{3}$, F by scale factor $\frac{1}{2}$ and G by scale factor $\frac{1}{4}$.

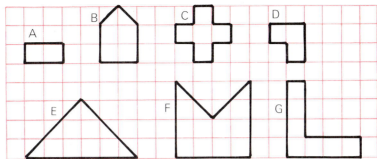

4 This is a view from above of an outdoor clothes drier.

  a $\triangle$ABC can be enlarged to fit $\triangle$DEF. OA = AD. What is the scale factor of the enlargement?

  b AB∥DE. Copy and complete:
    (i) $\angle$OAB = ...   (ii) $\angle$OAC = ...
    (iii) $\angle$BAC = ...

  c In the enlargement, name two other pairs of equal angles.

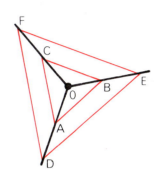

5 Is one shape an enlargement of the other in **a**? In **b**? Give reasons for your answers.

a
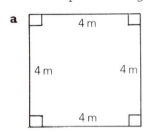

b
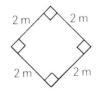
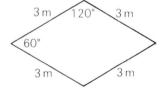

6 Why could a page of this book never be a reduction of a boxing ring?

7

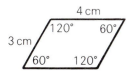

  a The trapezium and the parallelogram have equal angles, but neither is an enlargement nor a reduction of the other. Why not?

  b Sketch a trapezium which *is* a reduction of this trapezium.

  c Sketch a parallelogram which *is* an enlargement of this parallelogram.

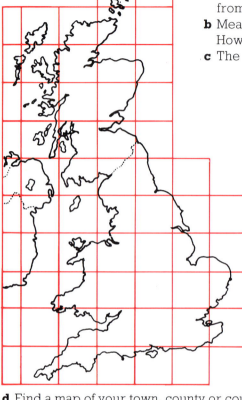

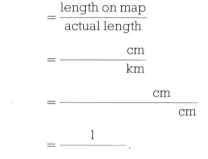

The map is *similar* to the actual shape of the land. You would see this outline from a spacecraft high above the earth.

a Each square is 100 km long. How many km from top to bottom of the grid?

b Measure this distance with your ruler. How many cm?

c The scale factor, or *scale*, of the map

$$= \frac{\text{length on map}}{\text{actual length}}$$

$$= \frac{\quad\quad\text{cm}}{\quad\quad\text{km}}$$

$$= \frac{\quad\quad\text{cm}}{\quad\quad\text{cm}}$$

$$= \frac{1}{\quad\quad}.$$

d Find a map of your town, county or country. Note the scale of the map. Measure the distance between two places on the map. Use the scale to convert this to the actual distance. Repeat this for different pairs of places, and make a table of distances between them.

SIMILAR SHAPES

Can you think of other everyday uses of the word 'similar'? Would you agree that to be similar, two people or things or ideas must have something in common?

In geometry, two shapes are *similar* if one is an enlargement (or reduction) of the other, that is if one is a scale version of the other. This means that:
  (i) corresponding angles are equal
  (ii) corresponding lengths are in the same ratio, that is have the same scale factor.

If one shape fits exactly on another, the shapes are *congruent*. The shapes are also similar, with scale factor 1.

Example 1

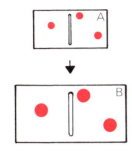

Two dominoes—are their outlines similar?
  (i) *Check angles.* All right angles, so corresponding angles are equal.
  (ii) *Check ratios of corresponding sides*

$$\frac{\text{Length of A}}{\text{Length of B}} = \frac{20\,\text{mm}}{30\,\text{mm}} = \frac{2}{3}.$$

$\dfrac{\text{Breadth of A}}{\text{Breadth of B}} = \dfrac{10\,\text{mm}}{15\,\text{mm}} = \dfrac{2}{3}.$ So equal ratios (scale factor $\frac{2}{3}$).

Their outlines are similar.

Example 2
The flat rectangular screens on the television sets are similar. The dimensions are in inches. Calculate x.

Ratio of heights $= \dfrac{15\,\text{inches}}{9\,\text{inches}} = \dfrac{5}{3}$. The scale factor is $\frac{5}{3}$.

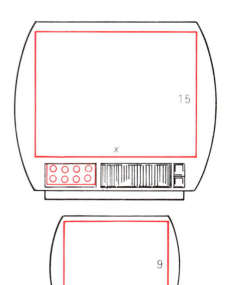

Method 1

$x = \dfrac{5}{3} \times 12 = 20$

Method 2

$\dfrac{x}{12} = \dfrac{5}{3}$

$3x = 60$ (Cross-multiplication)

$x = 20.$

SCALE FACTORS AND SIMILAR SHAPES

1 *Measure* the lengths and breadths of the windows and doors in **a** and **b**, and the fronts of **c** and **d**. By considering angles and scale factors (ratios of corresponding lengths) decide which pairs of outlines are similar.

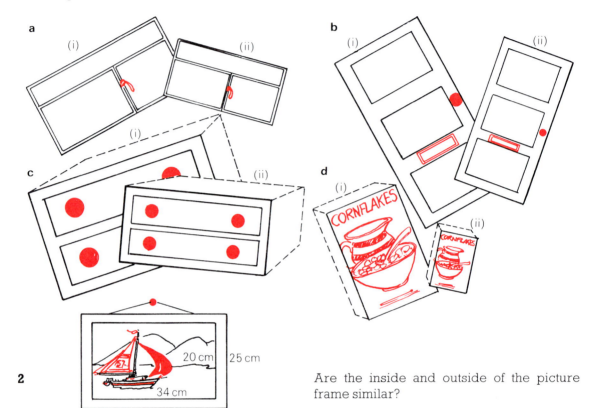

a (i) (ii)

b (i) (ii)

c (i) (ii)

d (i) CORNFLAKES (ii) CORNFLAKES

2

20 cm   25 cm

34 cm

40 cm

Are the inside and outside of the picture frame similar?

3 This toy car is $\frac{1}{6}$th life size.

a A real tyre has diameter 72 cm. What is the diameter of a tyre on the model?

b The rear wheel has twenty equally spaced spokes. What is the size of the angle between two spokes on:
(i) the real car   (ii) the model?

4

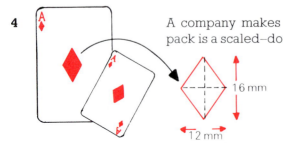

16 mm

12 mm

A company makes standard and junior playing cards. The junior pack is a scaled–down version, with scale factor 0·4.

a Calculate the lengths of the diagonals of the diamond (rhombus) at the centre of the junior ace.

b What can you say about the angles of the diamonds at the centres of the cards?

5 The flying ducks on Fenella's living room wall are similar.

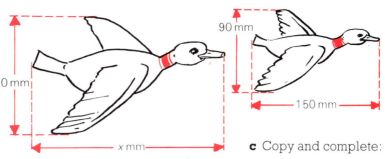

a Calculate: (i) x (ii) y. 120 mm
b The smallest duck's beak is 30 mm long. Calculate the lengths of the beaks of the other ducks.

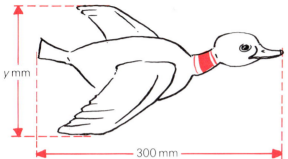

c Copy and complete:

Duck	Eye	Tail	Wing
Small	6 mm		
Medium		60 mm	
Large			84 mm

6 In the picture, the three cottages are similar to each other. Calculate:

a (i) x (ii) y
b the length AB
c the perimeter of the gable end of the smallest cottage.

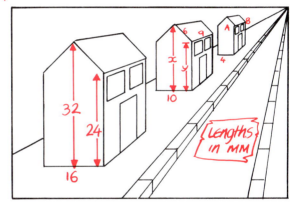

Lengths in MM

=== **Exercise 2B** ===

1

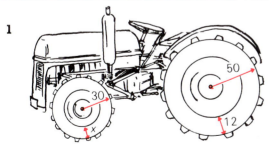

The tyres on this toy tractor are similar, and the lengths are in millimetres. Calculate the width of the smaller tyre (x).

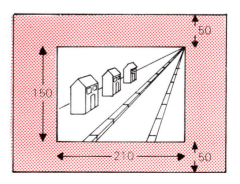

2 The outsides of the picture and the frame are similar. The lengths are in millimetres. Calculate:

a the height and breadth of the outside of the frame
b the width of the left and right-hand sides of the frame.

3 Which pair(s) does not consist of similar shapes? Why not?

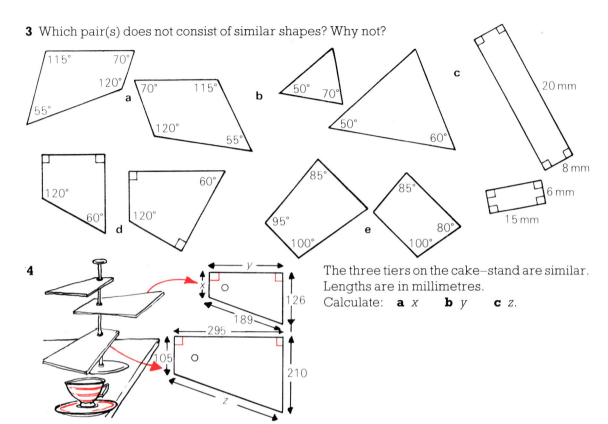

4 The three tiers on the cake–stand are similar. Lengths are in millimetres.
Calculate: **a** x **b** y **c** z.

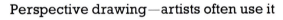

Perspective drawing—artists often use it

a Draw three shapes like those shown and choose a suitable point P.

b Draw lines, called *rays*, from each vertex to the point P.

c At some place nearer to P draw your shape again, making sure that corresponding sides are parallel and vertices lie on the rays.

d Rub out the edges of the solids you have drawn that cannot be seen.

e Thicken the remaining lines.

Investigate the connection between perspective drawing and similarity.

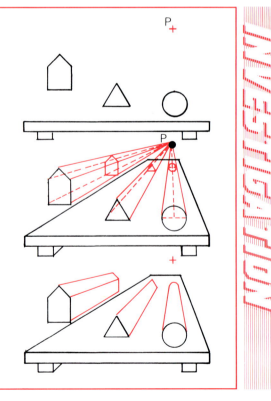

Television and cinema screens

=============================== Exercise 3 ===============================

1

Did you know? Every TV screen is similar to every other TV screen.

a Why is this?

b The height of every screen is $\frac{3}{4}$ of its width. Calculate:

(i) the height of a screen 24 cm wide

(ii) the width of a screen 24 cm high.

2 On squared paper draw a rectangular 'TV screen' 4 cm wide and 3 cm high. The *size* of a TV screen is usually quoted as the length of its diagonal.

a Measure the length of the diagonal of your rectangle. Check it by calculation.

b Calculate the width and height of a TV screen with diagonal of length:

(i) 10 cm   (ii) 45 cm   (iii) 55 cm.

3 Peggy makes a toy TV for her doll's house from a matchbox. It is a scaled-down version of her own TV which has a screen diagonal 50 cm long.

a The width of the toy TV screen is 20 mm. Calculate the height of the screen and the length of its diagonal.

b Calculate the scale factor used in making her toy TV.

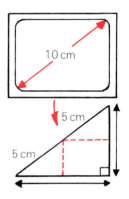

10 cm

5 cm

5 cm

4 Modern movies are made for wide screens which are not similar to TV screens. So there is a choice—

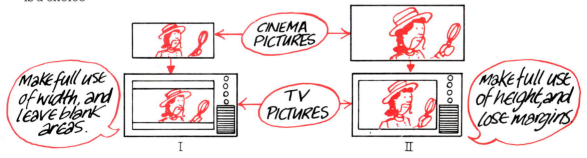

CINEMA PICTURES

Make full use of width, and leave blank areas.

TV PICTURES

Make full use of height, and lose margins.

I

II

The width of the cinema screen is twice its height. The pictures on the TV screen are placed centrally.

a In I, the TV screen has a 45 cm diagonal. Calculate the width of the TV picture, and the thickness of each black band.

b In II, the TV screen has a 55 cm diagonal. How much is cut off each side of the TV picture?

c Have you seen options I or II, or any other method, used?

The picture which a home computer 'sends' to the TV screen is 1024 units high and 1280 units wide. How could this best be fitted into a TV screen with a 30 cm diagonal?

TRIANGLES ARE SPECIAL

This tiling of congruent triangles contains three sets of parallel lines. Angles in corresponding positions in each pair of triangles are equal, so the triangles are *equiangular*.

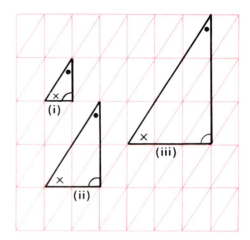

===== *Exercise 4* =====

1 Why are the sets of marked angles equal? The triangles are equiangular.

2 What is the ratio of corresponding sides in triangles:
   a (i) and (ii)   **b** (i) and (iii)   **c** (ii) and (iii)?

3 What is the scale factor of the enlargement from:
   a (i) to (ii)   **b** (i) to (iii)   **c** (ii) to (iii)?

4 a On squared paper draw a set of four equiangular triangles, not necessarily right–angled.
   b Check that corresponding sides in any pair have the same ratio.
   c Write down the scale factors of enlargement or reduction between pairs of triangles.

5 Can you find any equiangular triangles whose sides are not in the same ratio?

If two triangles are equiangular, their corresponding sides are in proportion. So the triangles are similar.
The converse is also true:
If two triangles have corresponding sides in proportion, they are equiangular. So the triangles are similar.

△s are **equiangular** ⟷ △s are **similar** ⟷ △s have lengths of **corresponding sides in proportion**

Note In △s ABC and DEF,

$\angle A = \angle D, \angle B = \angle E, \angle C = \angle F,$

and

$$\frac{AB}{DE} = \frac{BC}{EF} = \frac{AC}{DF}.$$

Corresponding sides are opposite equal angles.

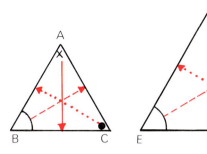

═══════════ *Exercise 5A* ═══════════

1 a Which pairs of triangles are equi-angular, and therefore similar?
  b Write down equal ratios of corresponding sides of the pairs of similar triangles. (Don't measure the sides—they are not accurate).

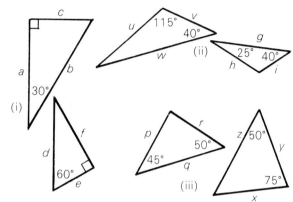

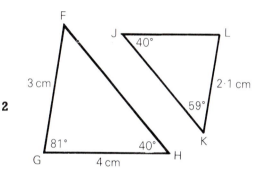

2

  a Explain why the triangles are similar.
  b Pick out the side corresponding to GH, and calculate its length correct to 1 decimal place.

3 Which pairs of triangles have their corresponding sides in proportion, and are therefore similar? (Arrange the sides in order of size, for example, 9, 12, 15 and 6, 8, 10. Then calculate ratios.)

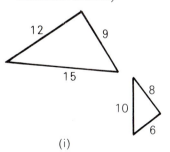

(i)

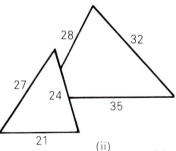

(ii)

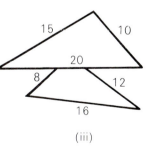

(iii)

4 a Jason proved that these two triangles are similar—can you?
  b He marked crosses at what he thought were equal angles. Sketch the two triangles, and mark pairs of equal angles. Did Jason blunder?

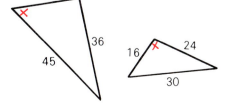

41

5 The edges of these shelves in the corner of the room are parallel.
  a Prove that △s ABC and PQR are similar.
  b If AB = 15 cm, calculate the length of PQ.

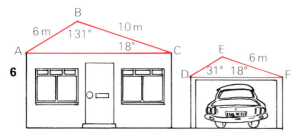

6

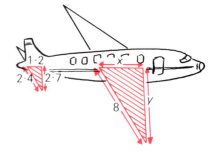

  a Prove that ends of the bungalow and garage roofs are similar.
  b Calculate the length of DE.

7 The three edges of the wing and tail are parallel.
  a Why are the shaded triangles similar?
  b Calculate: (i) x   (ii) y.
    All the lengths are in metres.

8

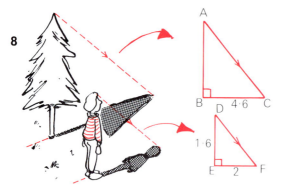

Both Tom and the tree cast long shadows in the sinking sun, making two similar triangles. Calculate the height of the tree. The lengths are in metres.

9 a Prove that the two triangles are similar.
  b Calculate the length of PQ, using:
    (i) similar triangles   (ii) trigonometry.

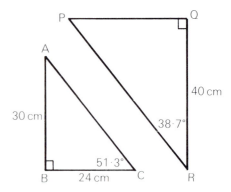

Calculate x, correct to 1 decimal place.

10

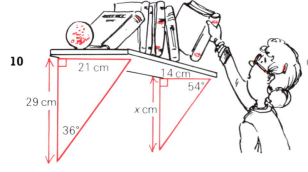

1 The chute in the playpark is right–angled at A.
  a Sketch the triangles in the picture, and mark the sizes of all the angles.
  b Copy and complete:
  △ABD is similar to △ . . . and to △ . . . .

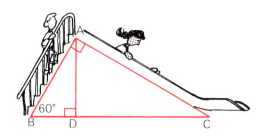

2

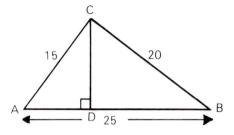

a Prove that:
  (i) ∠ACB = 90°
  (ii) all three triangles are equiangular.
b Find the scale factor connecting △s ABC and ADC.
c Calculate the length of:
  (i) AD   (ii) DB   (iii) CD.

3 Bluebell sets sail from A, 5 km North of Grimsby, for Spurn Head. B is its closest point to Grimsby.
  a Calculate the distance AS.
  b Use triangles ABG and ASG to prove that $\dfrac{x}{5} = \dfrac{5}{13}$.

  Now calculate x, correct to 2 decimal places.

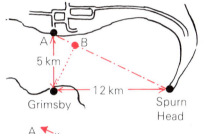

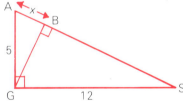

4

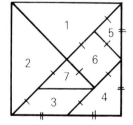

This square contains the seven tangram shapes. Prove that there are five similar triangles among these seven shapes.

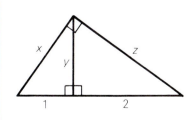

Calculate x, y and z, correct to 1 decimal place.

Parallel lines in triangles

DE is parallel to the base of △ABC. Check that, in each diagram, △s ABC and ADE are equiangular. So the triangles are similar.

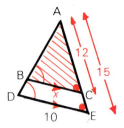

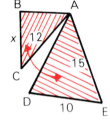

To calculate x, the reduction scale factor is $\dfrac{12}{15} = \dfrac{4}{5}$.

Method 1

$$x = \frac{4}{5} \times 10$$

$$= 8$$

Method 2

$$\frac{x}{10} = \frac{4}{5}$$

$$5x = 40$$

$$x = 8.$$

=================== **Exercise 6** ===================

1 Calculate x:

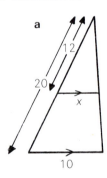

a

12

20

x

10

b

12

10

x

24

c

8

5

6

x

d

24

30

10

x

2 a Make a sketch of trapezium ABCD, and mark pairs of equal angles.
  b Name a pair of similar triangles.
  c Calculate the value of AP : PC.
  d AC = 12 cm. Calculate the lengths of AP and PC.

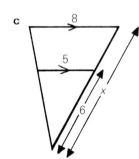

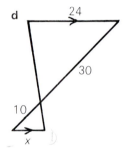

3

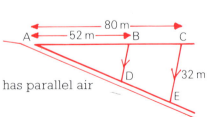

The Lucky Strike Gold Mine has parallel air shafts BD and CE.
  a Calculate the length of BD.
  b If AD = 54 m, calculate the length of DE, to the nearest metre.

4 DE and BC are horizontal. Calculate the
length of:
a AC   **b** EC.

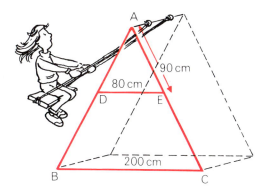

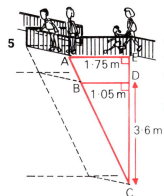

5 The observation platform at Corrieshalloch Gorge gives breathtaking
views. AE and BD are horizontal.
a Calculate the height CE of the platform above the bottom of the
gorge.
b If AC = 6·25 m, calculate the length of BC.

6

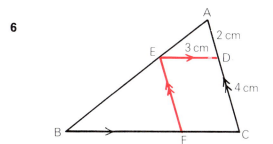

a Explain why DEFC is a parallelogram.
b Write down the lengths of EF and FC.
c Calculate the lengths of BC and BF.

7 The girder bridge spans the Middleton Canal. Calculate x.

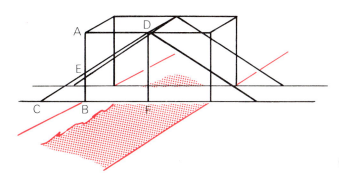

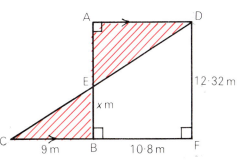

8

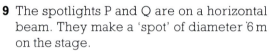

The ferryboat Faithful plies the 32 metres between the two slipways (A to B). Edna (E) and Dave (D) are in line with the ferry at F. Calculate the distance FB between the ferry and slipway B then.

9 The spotlights P and Q are on a horizontal beam. They make a 'spot' of diameter 6 m on the stage.

  a Write down the names of two similar triangles.

  b If PR = 17 m, calculate TR.

  c If Q is 12 m above the stage, how far is T above it?

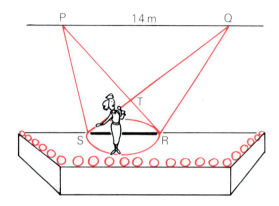

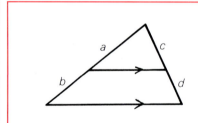

Using similar triangles, prove that $\dfrac{a}{b} = \dfrac{c}{d}$.

AREAS AND VOLUMES OF SIMILAR SHAPES

=== *Exercise 7* ===

1 a Why are all squares similar?

  b Copy and complete:

Enlargement from	(i) to (ii)	(i) to (iii)	(i) to (iv)
Linear scale factor	2 : 1 = 2		
Area scale factor	4 : 1 = 4		

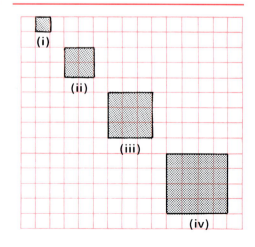

2 a The tiling is made of congruent triangles. Why are the shaded triangles similar?

b Copy and complete (count the small triangles for the areas):

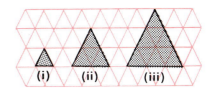

Enlargement from	(i) to (ii)	(i) to (iii)	(i) to (iv)		(i) to (*n*)
Linear scale factor					
Area scale factor					

3 Copy and complete this table for the 'nest' of similar hexagons.

Enlargement from	(i) to (ii)	(i) to (iii)	(i) to (iv)		(i) to (*n*)
Linear scale factor					
Area scale factor					

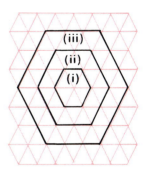

> If the linear scale factor for two similar shapes is n, then their area scale factor is n^2.

4

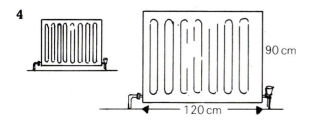

90 cm

120 cm

The radiators are similar, with linear scale factor 2. Calculate:

a the length and breadth of the smaller radiator

b the areas of the front faces of the radiators

c the area scale factor.

Does this agree with questions **1** and **2**?

5 The linear scale factor for these two similar tents is 3. Calculate:
  a the width and height of the front of the larger tent
  b the areas of the fronts of the tents
  c the area scale factor.
Check that the 'n^2 rule' holds.

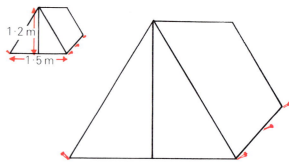

6

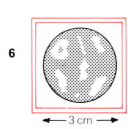

← 3 cm →

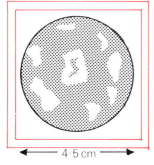

← 4·5 cm →

This is a photograph of the moon, and an enlargement of it. Calculate the area scale factor for the enlargement.

7 The enlargement scale factor is 3. Calculate the area of photographic paper needed for the enlargement.

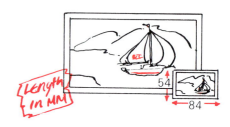

Length in MM

54

← 84 →

8

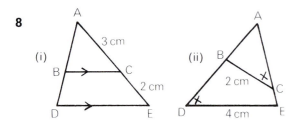

(i) A, 3 cm, B, C, 2 cm, D, E

(ii) A, B, 2 cm, C, D, 4 cm, E

a Prove that the triangles in each pair are similar.
b Write down the linear scale factor of the enlargement from △ABC to △ADE.
c If the area of △ABC is 5 cm², calculate the area of △ADE.

Exercise 8

1

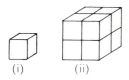
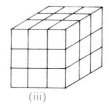

(i) (ii) (iii)

Each small cube has a 1 cm edge. Copy and complete:

Enlargement from	(i) to (ii)	(i) to (iii)	(i) to (iv)	. . .	(i) to (n)
Linear scale factor				. . .	
Volume scale factor				. . .	

2 Copy and complete this table for spheres ($V = \frac{4}{3}\pi r^3$. Don't work out $\frac{4}{3}\pi$.)

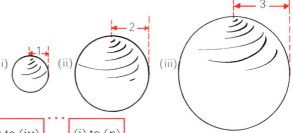

Enlargement from	(i) to (ii)	(i) to (iii)	(i) to (iv)	...	(i) to (n)
Linear scale factor	$2:1 = 2$			...	
Volume scale factor	*			...	

* $\frac{4}{3}\pi \times 2^3 \div \frac{4}{3}\pi \times 1^3 = 2^3 : 1^3 = 8:1 = 8$

> If the linear scale factor for two similar shapes is n, then their volume scale factor is n^3.

3 Perfume is sold in similar bottles of three different sizes. Corresponding edges of the bases of the bottles are 1 cm, 2 cm and 3 cm long, and the smallest bottle holds 2 millilitres of perfume. How much does each of the others hold?

4 These two cylindrical tins of paint are similar. Calculate the volume of paint in the larger tin.

8 cm   250 ml   16 cm   ? ml

5 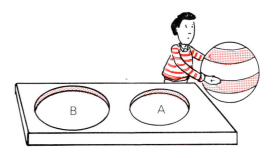 A model car is made using a scale factor of 1:50.

 a The model is 7·5 cm long. Calculate the length of the real car.
 b The area of the model's windscreen is 10 cm². Calculate the area of the windscreen on the real car in: (i) cm² (ii) m².
 c The car's petrol tank holds 50 litres. How much should the model's hold?

6 Two types of beach ball are tested by passing them through holes of the correct size. One ball has eight times the volume of the other, and just passes through a hole of diameter 30 cm. Calculate the diameter of the other hole.

7

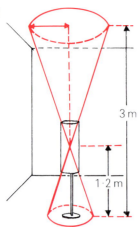

3 m

1·2 m

The standard lamp throws similar cones of light onto the floor and ceiling. Calculate the scale factor for:
a the heights of the cones
b the areas of the pools of light on the floor and ceiling
c the volumes of the two cones of light.

8 Star Cereals give a free sample packet (linear scale factor $\frac{2}{5}$) with every new large packet. The coloured design on the large packet covers 200 cm². Calculate:
a the area of the design on the free sample
b the volume of each packet
c the area of card needed for each.

9 Repeat question **8** for two cylindrical cartons of juice. The larger one has height 20 cm and diameter of base 10 cm, with a coloured design covering 80 cm² (take 3·14 for π).

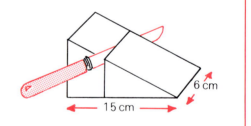

The sides of the wedge of cheese are right-angled triangles. The other faces are rectangles. A vertical cut is made in the cheese. Sketch the position of the cut which:
a halves the wedge (2 answers)
b quarters the wedge (4 answers).

6 cm

15 cm

Many photocopiers have an option button marked 71% . This allows you to print two A4 sheets onto one. Why is the button not marked 50% ?

A + B = B / A

a What would happen if you put a reduced sheet through again?
b Find out about other options, e.g. 62% , and investigate their meaning.

CHECK-UP ON **SCALE FACTORS AND SIMILAR SHAPES**

1 Which flag outline is not similar to the others? (Lengths in metres.)

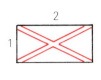

a

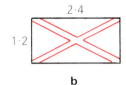

b

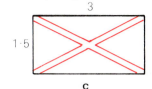

c

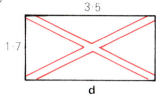

d

2 These two rectangular gates are similar.
The lengths are in metres. Calculate:
a the scale factor of the reduction
b x and y.

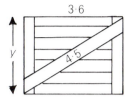

3 Prove that the shapes in each pair below are similar. (Lengths in centimetres.)

a

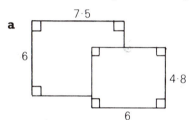

b

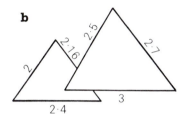

c

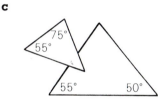

4 Calculate x in each drawing:

a

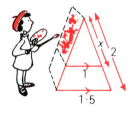

b

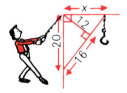

c

5 Calculate:
a x
b y.
(The lengths are in metres.)

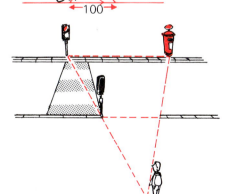

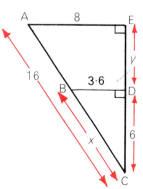

6 a One square has a side 1 cm long, another has a side 1 m long. Calculate the scale factor for
the enlargement of: (i) sides   (ii) areas.
b One cube has a side 1 mm long, another has a side 1 cm long. Calculate the scale factor for
the enlargement of: (i) sides   (ii) volumes.

Investigation

Middle Wallop Youth Group are having a Treasure Hunt. Part of the football field is lined off in 10 metre squares.

The treasure (a voucher for sports goods) is buried at one of the crossing points. The members are given these two clues to help them to find the treasure:

$y = x$ $y = -x + 6$

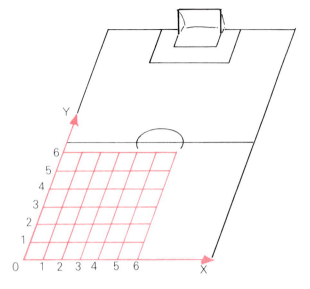

1 How many crossing points are there?

2 Copy the grid on squared paper. Imagine you are taking part in the Treasure Hunt. Can you find the treasure, using the two clues?

3 If you have found a method that works, try it for these pairs of clues:
   a $y = x$, $y = -x + 4$ **b** $y = 5$, $x + y = 8$ **c** $y = 2x$, $y = x + 2$.

4 Can you think of another way to find the treasure?

STRAIGHT LINE GRAPHS AGAIN

The first clue is $y = x$. Possible locations of the treasure at the crossing points are:

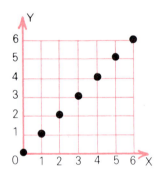

x	0	1	2	3	4	5	6
y	0	1	2	3	4	5	6

The locations are marked by the seven dots. If the treasure had been hidden *anywhere* on the straight line through the points, John and Alison and the others would have to search the whole of the line from $x = 0$ to $x = 6$, that is for $0 \leqslant x \leqslant 6$.

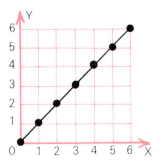

Possible locations for the clue $y = -x + 6$ are:

x	0	1	2	3	4	5	6
y	6	5	4	3	2	1	0

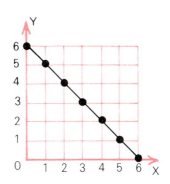

All crossing points on the line are also shown on the graph.
Taken together, the clues lead to the actual position of the treasure at the point (3, 3).

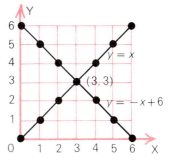

=========================== *Exercise 1* ===========================

For each pair of clues:
  (i) Make a table of values.
  (ii) Plot the points on squared paper.
  (iii) Draw the straight lines through the points.
  (iv) Find the point where the treasure is hidden.

1A $y = x + 2$ and $y = -x$, for $-3 \leqslant x \leqslant 3$.
2A $y = x$ and $y = -x + 4$, for $-4 \leqslant x \leqslant 2$.
3A $y = 2x$ and $y = -x - 3$, for $-4 \leqslant x \leqslant 2$.
4A $y = x + 2$ and $y = 8 - x$, for $-2 \leqslant x \leqslant 8$.
5A $x + y = 8$ and $x - y = 4$, for $0 \leqslant x \leqslant 8$.
6B $2x + y = 6$ and $x - 2y = 8$, for $0 \leqslant x \leqslant 8$.
7B $y = x + 10$ and $y = 1 - 2x$, for $-6 \leqslant x \leqslant 2$.
8B $y = 2x + 1$ and $y = 3x - 1$, for $-2 \leqslant x \leqslant 6$.

What can you say about the person who buried the treasure and gave the teams the clues $y = 2x$ and $y = 2x + 3$?

SOLVING A PAIR OF LINEAR EQUATIONS GRAPHICALLY

Draw the graph of each equation. The point of intersection of the graphs gives the solution of the pair of equations.

Example

Solve the pair of equations $\left.\begin{array}{l}3x+2y=12\\3x-2y=0\end{array}\right\}$.

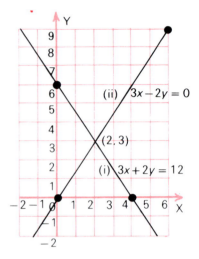

In Chapter 1 you saw that you only have to plot two points to draw the graph of a straight line. Choose the points where the line crosses the x- and y-axes, if possible.

(i) For $3x+2y=12$:
   When $x=0$, $2y=12$ so $y=6$, giving the point $(0,6)$.
   When $y=0$, $3x=12$ so $x=4$, giving the point $(4,0)$.

(ii) For $3x-2y=0$:
   When $x=0$, $y=0$, giving $(0,0)$.
   When $x=6$, $18-2y=0$ so $y=9$, giving the point $(6,9)$.
   The graphs cross at the point $(2,3)$.

So the solution of the pair of equations is $x=2$, $y=3$, or $(2,3)$.

Note When taken together, $3x+2y=12$ and $3x-2y=0$ are sometimes called *simultaneous equations*. Why do you think this is?

Exercise 2

By drawing the graphs of the equations, solve each pair of equations:

1A $\left.\begin{array}{l}x+y=6\\x-y=4\end{array}\right\}$ **2A** $\left.\begin{array}{l}x+2y=6\\x-y=3\end{array}\right\}$ **3A** $\left.\begin{array}{l}y=x+3\\y=-x-3\end{array}\right\}$

4A $\left.\begin{array}{l}x+3y=3\\x-3y=9\end{array}\right\}$ **5A** $\left.\begin{array}{l}y=2x+4\\y=-x-2\end{array}\right\}$ **6A** $\left.\begin{array}{l}3x-y=7\\y=5\end{array}\right\}$

7B $\left.\begin{array}{l}2x-y=0\\3x+3y=-9\end{array}\right\}$ **8B** $\left.\begin{array}{l}x+y=0\\x-y=0\end{array}\right\}$ **9B** $\left.\begin{array}{l}y=x\\y=7-x\end{array}\right\}$

Use 2 mm squared paper for **10B**, **11B** and **12B**:

10B $\left.\begin{array}{l}x+y=2\\2x-4y=1\end{array}\right\}$ **11B** $\left.\begin{array}{l}2x+3=0\\3x-y+1=0\end{array}\right\}$ **12B** $\left.\begin{array}{l}x+y=6\\5y-5x=6\end{array}\right\}$

PAIRS OF EQUATIONS

1A a Bart jumps, followed by Adele—how many seconds later?
b One passes the other on the way down. Who, when and at what height?

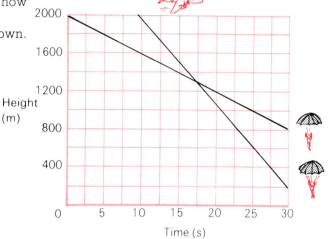

2A

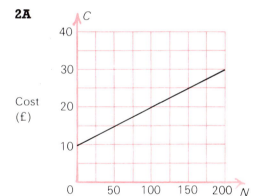

a North Gas Board has a standing charge of £10 per quarter, plus 10p per unit.
Using the graph, copy and complete:

Number of units (N)	0	50	100	150	200
Cost (£C)	10				

b South Gas Board has no standing charge, but each unit costs 20p. Copy and complete:

Number of units (N)	0	50	100	150	200
Cost (£C)	0	10			

c Draw graphs of C against N for both Boards on the same page.
d Where the two graphs cross, the Boards charge the same.
   (i) How much?   (ii) For how many units?
e Lynn Taylor uses 150 units. Which Board would be cheaper for her?
f (i) Copy and complete: The equation of the first graph is $C = \frac{1}{10}N + \ldots$.
   (ii) Write down the equation of the second graph, $C = \ldots$.

3A a 'Maximiles' hire out motor bikes. For a Lightning Streak model the terms are £20 deposit and a daily charge of £5.
  (i) Using these scales, draw a graph of the cost over 5 days.
  (ii) Write down an equation for the cost of hire, £C, for N days.
b 'Speedwheels' also hire out motor bikes. Their charge is £10 a day, without any deposit.
  (i) Draw the Speedwheels graph on the same page.
  (ii) Write down an equation for C in terms of N.
c Which method is cheaper for 3 days? 4 days? 5 days?

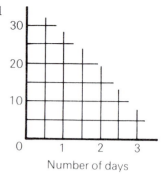

4B The cost of producing a new computer game, *Blast-off 2020*, is £200 plus £15 per cassette.

a Copy and complete the equation for the cost (£y) of producing x cassettes: $y = \ldots + \ldots$.
Each cassette is sold for £25.
b Copy and complete the equation for the money got (£y) by selling x cassettes: $y = \ldots$.
c Draw the straight line graphs of these two equations on the same squared paper for $0 \leqslant x \leqslant 50$. Scales: horizontally 5 units to 1 square, vertically 100 units to 1 square.
d Investigate the 'break–even' value for x, and use the graphs to describe the profit and loss story for the production of *Blast-off 2020*.

OFF THE PAGE—SOLVING A PAIR OF EQUATIONS BY SUBSTITUTION

Alison and John had finished the *Blast-off* investigation. 'What next?' they asked Mrs O'Grady.
'Make up a pair of equations for yourselves', she said.
They ran into trouble with $y = 2x$ and $y = x + 5$—the lines didn't meet on the graph paper.
'No solution?' said John.
'Let's see', said Alison. 'Where the graphs meet the y–values are the same, so you really don't need a graph.'

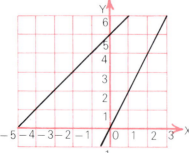

$$\left.\begin{array}{l} \boldsymbol{y} = x + 5 \\ \boldsymbol{y} = 2x \end{array}\right\}$$

So $2x = x + 5 \ldots\ldots\ldots$ (i)

    $x = 5$

and $y = 2x = 10$.

The point of intersection of the lines is (5, 10), and this is the solution of the pair of equations.

Note   In (i), $2x$ is *substituted* for y, so the method is called *substitution*.

Solve these pairs of equations by substitution:

1A $\left.\begin{array}{l} y = 2x \\ y = x+10 \end{array}\right\}$ **2A** $\left.\begin{array}{l} y = 3x \\ y = 20-x \end{array}\right\}$ **3A** $\left.\begin{array}{l} y = 2x-8 \\ y = x+1 \end{array}\right\}$

4A $\left.\begin{array}{l} y = 4x+12 \\ y = x+12 \end{array}\right\}$ **5A** $\left.\begin{array}{l} y = 1-2x \\ y = x+10 \end{array}\right\}$ **6A** $\left.\begin{array}{l} y = 3x+1 \\ y = 6-2x \end{array}\right\}$

7B $\left.\begin{array}{l} y = \frac{1}{2}x \\ y = x+8 \end{array}\right\}$ **8B** $\left.\begin{array}{l} y = 8x \\ y = 4x+1 \end{array}\right\}$ **9B** $\left.\begin{array}{l} y = 3-x \\ y = 3x+4 \end{array}\right\}$

10B $\left.\begin{array}{l} y = 2x \\ 3x+y = 5 \end{array}\right\}$ **11B** $\left.\begin{array}{l} y = 2x-3 \\ x+y = 3 \end{array}\right\}$ **12B** $\left.\begin{array}{l} 2x+y = 8 \\ x-y = 1 \end{array}\right\}$

13B $\left.\begin{array}{l} 4x+2y = 0 \\ 5x+2y = -1 \end{array}\right\}$ **14B** $\left.\begin{array}{l} 2x+2y = 7 \\ 3x+4y = 12 \end{array}\right\}$ **15B** $\left.\begin{array}{l} 4x+3y = 7 \\ 2x+2y = 3 \end{array}\right\}$

SOLVING A PAIR OF EQUATIONS BY ELIMINATION

A puzzle

The sum of two numbers is 20. Their difference is 5. Find the numbers.

Alison's solution

If the numbers are x and y,

$$x + \mathbf{y} = 20$$
$$x - \mathbf{y} = 5$$

$Add \quad 2x = 25 \ldots\ldots\ldots$ (i)

$$x = 12\frac{1}{2}$$

Put $x = 12\frac{1}{2}$ in the first equation: $12\frac{1}{2} + y = 20$

$$y = 20 - 12\frac{1}{2} = 7\frac{1}{2}$$

The numbers are $12\frac{1}{2}$ and $7\frac{1}{2}$.

Check $12\frac{1}{2} + 7\frac{1}{2} = 20$. $12\frac{1}{2} - 7\frac{1}{2} = 5$.

Note When the two equations are added, y is *eliminated* in (i), so the method is called *elimination*.

Exercise 5

Solve these pairs of equations by adding or subtracting them to eliminate x or y, etc.

1A $\left.\begin{array}{l} x+y = 12 \\ x-y = 6 \end{array}\right\}$ **2A** $\left.\begin{array}{l} x+y = 15 \\ x-y = 5 \end{array}\right\}$ **3A** $\left.\begin{array}{l} x+y = 10 \\ x-y = 3 \end{array}\right\}$

4A $\begin{array}{l} a+3b = 3 \\ a-3b = 9 \end{array}$ **5A** $\left.\begin{array}{l} c+4d = 22 \\ c-4d = -18 \end{array}\right\}$ **6A** $\left.\begin{array}{l} 2e+3f = 9 \\ 2e-3f = 15 \end{array}\right\}$

7A $\left.\begin{array}{l} 3p-q = 8 \\ 3p+q = 4 \end{array}\right\}$ **8A** $\left.\begin{array}{l} 2p+q = -2 \\ 4p-q = -1 \end{array}\right\}$ **9A** $\left.\begin{array}{l} u+2v = 2 \\ u-2v = -8 \end{array}\right\}$

10A $\begin{aligned}2x+y&=17\\x+y&=9\end{aligned}\Big\}$ (*Subtract!*)   **11A** $\begin{aligned}5x+y&=4\\2x+y&=1\end{aligned}\Big\}$ **12A** $\begin{aligned}3x+3y&=3\\x+3y&=3\end{aligned}\Big\}$

13A $\begin{aligned}2a-b&=11\\a-b&=5\end{aligned}\Big\}$ **14A** $\begin{aligned}5m-2n&=13\\m-2n&=1\end{aligned}\Big\}$ **15A** $\begin{aligned}7u-3v&=10\\2u-3v&=5\end{aligned}\Big\}$

16B $\begin{aligned}5x-2y&=11\\x-2y&=-1\end{aligned}\Big\}$ **17B** $\begin{aligned}3p-5q&=-1\\p-5q&=3\end{aligned}\Big\}$ **18B** $\begin{aligned}4s-3t&=15\\-2s-3t&=-3\end{aligned}\Big\}$

Another Puzzle

When do you add the equations, and when do you subtract them?

A challenge

John always looks for difficulties. 'What about these?' he asks. 'Your method won't work for them.'

$$\begin{aligned}x+2y&=-3\\3x+y&=1\end{aligned}\Big\}\quad \textit{Add: } 4x+3y=-2 \quad \textit{Subtract: } -2x+y=-4.$$

John is right. Neither x nor y is eliminated.

Alison's solution

$x+2y=-3$	$\times\mathbf{1}$	$x+\mathbf{2y}=-3$	**or**	$x+2y=-3$	$\times\mathbf{1}$	$x+\mathbf{2y}=-3$
$3x+y=1$	$\times\mathbf{2}$	$6x+\mathbf{2y}=2$		$3x+y=1$	$\times\mathbf{-2}$	$-6x-\mathbf{2y}=-2$
	Subtract	$\overline{-5x=-5}$			*Add*	$\overline{-5x=-5}$
		$x=1$				$x=1$

Put $x=1$ in the first equation: $\begin{aligned}1+2y&=-3\\2y&=-4\\y&=-2\end{aligned}$

The solution is $(1, -2)$.

Remember!

An equation is like a balance. You must do the same to **each side**—add, subtract, multiply or divide.

Exercise 6

Solve these pairs of equations:

1A $\begin{aligned}x+2y&=6\\2x-y&=7\end{aligned}\Big\}$ **2A** $\begin{aligned}x+3y&=1\\2x-y&=2\end{aligned}\Big\}$ **3A** $\begin{aligned}5x+y&=4\\x-2y&=3\end{aligned}\Big\}$

4A $\begin{aligned}3a+2b&=4\\a+b&=1\end{aligned}\Big\}$ **5A** $\begin{aligned}4c+2d&=2\\c+d&=0\end{aligned}\Big\}$ **6A** $\begin{aligned}5e+f&=1\\e+2f&=2\end{aligned}\Big\}$

7A $\begin{aligned}3x-4y&=14\\x+y&=0\end{aligned}\Big\}$ **8A** $\begin{aligned}5x-2y&=7\\x-3y&=4\end{aligned}\Big\}$ **9A** $\begin{aligned}x-y&=1\\3x-2y&=4\end{aligned}\Big\}$

10B $\left.\begin{array}{l}3x+2y=15 \\ 5x-3y=25\end{array}\right\}\begin{array}{l}\times 3 \\ \times 2\end{array}$

11B $\left.\begin{array}{l}2x-3y=7 \\ 7x+2y=12\end{array}\right\}$

12B $\left.\begin{array}{l}4x+2y=-6 \\ x-4y=3\end{array}\right\}$

13B $\left.\begin{array}{l}4a-3b=0 \\ 5a-4b=-1\end{array}\right\}$

14B $\left.\begin{array}{l}2c-3d=12 \\ 3c-2d=13\end{array}\right\}$

15B $\left.\begin{array}{l}2e+3f=-5 \\ 3e+2f=0\end{array}\right\}$

16B $\left.\begin{array}{l}3x+4y-20=0 \\ 4x-3y-10=0\end{array}\right\}$

17B $\left.\begin{array}{l}11x+3y+7=0 \\ 2x+5y-21=0\end{array}\right\}$

18B $\left.\begin{array}{l}3x+5y-23=0 \\ 5x+2y-13=0\end{array}\right\}$

USING THE EQUATION OF THE GRAPH OF A STRAIGHT LINE $y=ax+b$

A free–fall parachutist calls out the distance d metres he falls from a marked position after time t seconds.

t	3	6	9	12
d	25	40	55	70

The graph of d against t is a straight line. So they are connected by an equation of the form '$y=ax+b$'.

Here $d = \boldsymbol{at}+\boldsymbol{b}$ is a *formula* for d.

When $t=3$, $d=25$: so $\quad 25=3a+b$.
When $t=6$, $d=40$: so $\quad 40=6a+b$
$$\text{Subtract} \quad -15=-3a$$
$$a=5.$$
Put $a=5$ in the first equation: $25=15+b$
$$b=10.$$
The formula is $d=5t+10$.
Check. When $t=12$, $d=70$.

Note Using the **graph or formula**, when $t=7$, $d=45$.
Using the **formula**, when $t=20$, $d=110$.

Distance (m)

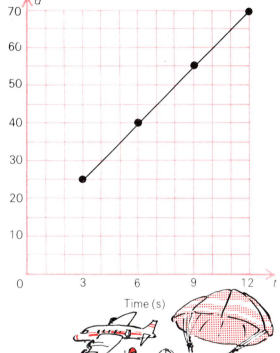

Time (s)

========== *Exercise 7* ==========

1A In an experiment, measurements of p and t give a straight line graph of p against t. So $p=at+b$.
When $p=11$, $t=3$ and when $p=3$, $t=1$. Calculate a and b, and write down the formula for p.

2A In another experiment, the graph of Y against X is a straight line, so $Y=aX+b$.
When $X=10$, $Y=32$ and when $X=20$, $Y=62$. Calculate a and b, and write down a formula for Y.

3A The formula for the pressure P at depth x in the ocean is of the form $P = ax + b$. At the surface (depth $= 0$) $P = 15$, and when $x = 60$, $P = 45$.

    a Calculate a and b, and write down the formula for P.

    b Draw a graph for $0 \leqslant x \leqslant 100$.

    c Estimate the pressure when $x = 80$, using: (i) the graph (ii) the formula.

    d Why is $P = 15$ at the surface?

4A The school's Young Enterprise Group decide to produce scribble pads. Before they start to produce, there is a basic cost of £a. Thereafter the cost of producing each pad is b pence.

    a Prove that the formula for the total cost C pence for n pads is $C = 100a + nb$.

    b For 100 pads, $C = 1400$ and for 50 pads $C = 900$. Find the Group's working formula for C.

    c Calculate the cost of producing 250 pads.

5B Water boils at $212°F$ and $100°C$, and freezes at $32°F$ and $0°C$. The formula connecting $F°$ Fahrenheit and $C°$ Celsius is $F = aC + b$.

    a Calculate a and b, and write down the formula for F.

    b Draw a graph for $-10 \leqslant C \leqslant 110$.

    c Estimate F when $C = 40$, using: (i) the graph (ii) the formula.

---

    a Discuss the solution of the *system of equations*:

        $x + 2y = 8$, $3x + y = 9$, $2x - y = 1$ and $x - y = -1$.

        Include the graphs of the equations on the same diagram in your answer.

    b How many points of intersection are possible for four straight lines in general?

Investigate the solutions of these pairs of equations.

Try different methods: for example, 'trial and error', tables of values, graphs, substitution.

1 $\left. \begin{array}{l} y = x \\ y = x^2 \end{array} \right\}$ **2** $\left. \begin{array}{l} y = x \\ y = x^3 \end{array} \right\}$ **3** $\begin{array}{l} y = x \\ y = \dfrac{16}{x} \end{array}$ **4** $\left. \begin{array}{l} y = x \\ x^2 + y^2 = 8 \end{array} \right\}$

PROBLEM SOLVING

A problem

Total cost £15

Total cost £10

Cost of each?

Introduce letters

Cost £x

Cost £y

Write equations

$$3x + y = 15 \quad \times \mathbf{2} \qquad 6x + 2y = 30$$
$$x + 2y = 10 \quad \times \mathbf{1} \qquad x + 2y = 10$$

$$\text{Subtract} \qquad 5x = 20$$
$$x = 4$$

Put $x = 4$ in the first equation: $\quad 12 + y = 15$
$$y = 3$$

Solve the problem

Cost is £4

Cost is £3

Check

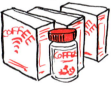
Total cost is
$3 \times £4 + 1 \times £3$
$= £15$

Total cost is
$1 \times £4 + 2 \times £3$
$= £10$

Exercise 8

1A

Total cost £9.

Total cost £12.

Cost of each?

2A

Total playing time 13 hours.

Total playing time 22 hours.

Playing time of each?

61

3A

Total length 30 m.

Total length 28 m.

Length of each?

4A

Total cost £8.

Total cost £7.

Cost of each?

5A

Hiring cost £12.
(Watch signs here!)

Hiring cost £3.

Hiring cost of this lot?

6A

Cost £8.

Cost £1.

Total cost?

7A

Cost 70p.

Cost £1·30.

Total cost?

8A

Cost £1·60.

Cost £1·70.

Total cost?

9B In Australian football, a goal scored below the cross–bar is given a certain number of points, and a goal scored above the cross–bar is given a different number of points.

West Town score 2 below and 5 above the bar.

East Town score 3 below and 2 above.

The final score is 'West Town 17, East Town 20'.

a What is the scoring system?

b The return game is drawn, 7 points each. How was this score made?

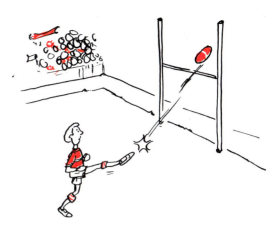

10B  Mr and Mrs Yarrow's Building Society will lend them $2\frac{1}{2}$ times their main salary plus the amount of the other salary to buy a house. If they can borrow £33 000, and the difference between their two salaries is £2000, find how much each earns.

11B A rectangular park is x metres long and y metres broad. The difference between the length and breadth is 50 m, and the perimeter of the park is 200 m. Calculate its length and breadth.

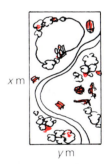

x m

y m

12B A machine can make 40 microchips per hour, and an older machine can make 30 per hour. An order for 400 has to be met in exactly 12 hours of machine time.

a Find the number of hours that each machine will have to operate.

b What is the shortest number of hours in which the two machines together could meet an order of 400 microchips?

In the 'Bonus and Penalty' game there are two types of token:

Bonus tokens

Penalty tokens

Alex scored 9 points with these tokens:

Jean scored 17 points with these tokens:

1 What is each type of token worth?

2 Investigate the least number of tokens needed to make scores of 1, 2, 3, 4, . . . , 12.

PAIRS OF EQUATIONS

1 *Can you find the treasure?*

Clues: 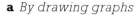 $y = -x + 5$ $y = x + 1$

(i) Make a table for each clue, taking $x = -2, -1, 0, 1, 2, 3, 4, 5$.

(ii) Plot the two sets of points.

(iii) Draw the straight lines through the points.

(iv) Find where the treasure is hidden.

2 *Can you solve these pairs of equations?*

a *By drawing graphs*

(i) $\begin{aligned} y &= x+1 \\ x+2y &= 8 \end{aligned}$

(ii) $\begin{aligned} y &= 2x \\ 2y &= x+6 \end{aligned}$

b *By substitution*

(i) $\begin{aligned} y &= 5x \\ y &= 3x-6 \end{aligned}$

(ii) $\begin{aligned} y &= 3x \\ x+y+8 &= 0 \end{aligned}$

c *By eliminating x or y*

(i) $\begin{aligned} 3x+2y &= 11 \\ x-2y &= 9 \end{aligned}$

(ii) $\begin{aligned} 3x+5y &= 11 \\ 2x+4y &= 9 \end{aligned}$

3 *Can you find and use a formula?*

On its first day of flowering a Christmas cactus plant had three flowers. Two more flowers opened the next day. The total number of flowers, f, after n days is given by the formula $f = an+b$, where a and b are numbers.

a (i) Check that for $n = 1$ and $f = 3$, $a+b = 3$.

(ii) Make another equation involving a and b.

(iii) Solve the two equations for a and b.

(iv) Complete the formula $f = \ldots n + \ldots$.

b If the plant first flowered on 23rd December, how many flowers did it have on 1st January (assuming that none had withered)?

4 *Can you solve this problem?*

Inside the nucleus of an atom there are protons and neutrons. Inside these there are 'Upquarks' and 'Downquarks'. Two 'Ups' and one 'Down' make a proton. One 'Up' and two 'Downs' make a neutron. The electrical charges of protons and neutrons are 1 and 0, respectively.

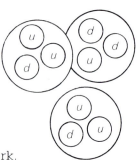

a Call the charge on an Up u, and on a Down d.

Make two equations involving u and d.

b Solve the equations to find the electrical charge on each type of quark.

USING BANKS, BUILDING SOCIETIES AND NATIONAL SAVINGS

(i) Bank accounts

Billy O'Reilly decides to open a current account in the Moneymaker Bank. The bank offers all of these services. Do you know what they mean?

Billy pays in £50, and is given a cheque book. Here is his first cheque:

Cheque stub →

Cheque number   Bank code   Account number

← Signature

═══════════════════ *Exercise 1* ═══════════════════

1A Make a copy of Billy's second cheque (number 00002) for £12·50, payable to 'Hot Hi-fi', on 28.10.88.

Every month a *bank statement* comes through Billy's letter box. It shows all the money paid in (credits) and paid out (debits), and the balance.

MONEYMAKER BANK STATEMENT		ACCOUNT NUMBER		SHEET NO.	
WILLIAM O'REILLY		2020202		1	

DATE	CODE	DESCRIPTION	DEBITS	CREDITS	BALANCE
20 OCT 88	CSH			50·00	50·00
25 OCT 88	CHQ	00001 TOP TEN	25·10		24·90
28 OCT 88	CHQ	00002 HOT HIFI	12·50		12·40
30 OCT 88	CHQ	REGIONAL COUNCIL		475·00	487·40
5 NOV 88	CHQ	00003	100·00		

Cash
Cheque
Cheque number

2A a Explain what the entries mean on:
  (i) 20 Oct   (ii) 25 Oct   (iii) 30 Oct.
b How much did Billy have in his account on:
  (i) 25 Oct   (ii) 30 Oct   (iii) 5 Nov?
c On 9 November he paid the Dodgy Gas Co £35·70 by cheque.
  How would this transaction appear in his bank statement?

3A Billy soon realised that the bank did not pay any interest on balances in *current accounts*. So he opened a *deposit account* which offered interest at 7% p.a. (per annum = yearly). How much interest would he receive in one year on:
a £200 **b** £700 **c** £1250 **d** £35?

(ii) Building Society accounts

4A

Building Societies also want you to invest money with them. Billy's sister Maeve invests £960 with the Mushroom Building Society for 3 months. Copy and complete:

a Interest for 1 year = 8% of £960 = ...
b Interest for 3 months = ...

5A Jim Wilson, a pensioner, puts his savings of £6600 into the Money Mountain Building Society. He takes out the interest at the end of each year. He uses it to buy Premium Bonds at £5 each, hoping to win the Big Prize. How many Bonds will he have by the end of the second year?

6A The Union Jack Building Society's Ordinary Account graph:

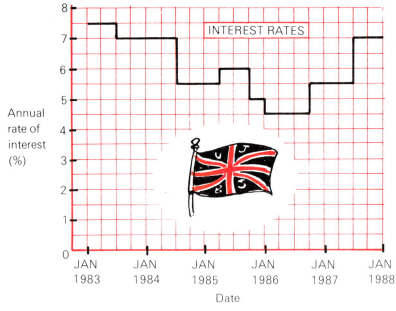

Annual rate of interest (%)

INTEREST RATES

Date

a How often has the rate changed during the five years?

b Calculate the interest on:
(i) £3000 for the first six months of 1984
(ii) £1200 for the first nine months of 1986
(iii) £1500 for the whole of 1987 (Careful!).

7A

Make your money

GR0W!!

Invest it in Mushroom Building Society

Invest	£1000	£10 000	£20 000	£50 000
After 1 year	£1073	£10 800	£21 650	£54 225

a Calculate the rate % p.a. for each deposit.
b Write a sentence about the Society's policy for attracting investments.

(iii) National Savings

Joshua was browsing in the Post Office one day. He discovered that the National Savings Investment Account pays 12% p.a., before tax. (One month's notice of withdrawal needed.) Here is his account a year later:

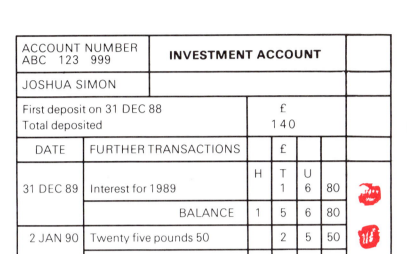

ACCOUNT NUMBER ABC 123 999	INVESTMENT ACCOUNT					
JOSHUA SIMON						
First deposit on 31 DEC 88 Total deposited			£ 1 4 0			
DATE	FURTHER TRANSACTIONS		£			
		H	T	U		
31 DEC 89	Interest for 1989		1	6	80	
	BALANCE	1	5	6	80	
2 JAN 90	Twenty five pounds 50		2	5	50	
	BALANCE					

8A a How much is Joshua's:
 (i) first deposit
 (ii) first year's interest?
b Check the interest.
c Calculate the balance on 2 Jan 90.
d Why did Joshua choose the Investment Account rather than the Ordinary Account?

9A Mrs Dixon decides to buy £100 of National Savings Certificates.
 a How much will they be worth in:
 (i) 1 year (ii) 3 years (iii) 5 years?
 b Why should she keep them for the full five years if possible?

National Savings Certificates
£25 units increase like this:

After	Value	Annual Interest
1 year	£26·63	6·52%
2 years	£28·63	7·51%
3 years	£31·11	8·66%
4 years	£34·19	9·90%
5 years	£38·03	11·23%

10A

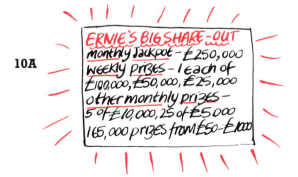

ERNIE'S BIG SHARE-OUT
monthly jackpot – £250,000
weekly prizes – 1 each of
£100,000, £50,000, £25,000
other monthly prizes –
5 of £10,000, 25 of £5,000
165,000 prizes from £50–£1000

Mr Dixon buys £10 of *Premium Bonds*. No interest, but......?
 a Why do you think he has a flutter on Premium Bonds?
 b What disadvantage have they?
 c What are the biggest and smallest prizes?
 d What do the initials ERNIE stand for?

11B Interest on Investment Accounts, Ordinary Accounts (after the first £70), Income Bonds and Deposit Bonds is liable to income tax at 27p in the £. The following rates are before tax. Calculate them after deduction of tax at 27p in the £:
 a 8% **b** 11% **c** 12·5%.

Investigate the ways in which you can invest any money you have saved. Banks have various offers to attract new customers, and Building Societies, National Savings, Insurance Companies, property, shares and unit trusts are other possibilities. Some deduct income tax, some don't; some are risky, some are not; some will be rewarding, some may not be.

BORROWING MONEY—LOANS

Tony Ferrara wanders round the showroom. He has decided to buy on the 'never never', Hire Purchase. This way he'll not need to use his savings.

Moneymaker Bank give him this table of loan repayments.

24 MONTHLY PAYMENTS			
Amount of loan	Monthly payment	Total amount payable	Total interest
£500	£25·00	£600	£100
1000	50·00		
2000	100·00		
3000	150·00		

=== *Exercise 2* ===

1A a Copy and complete the table above.
  b If he decides to buy a Maxi for £3000 how much will Tony:
     (i) pay monthly for a full loan   (ii) pay back *more* than he borrowed?

2A However, after doing his sums, Tony finds that he can only afford repayments of £75 monthly. Which of the three cars can he buy?

3A

Anna Santiago's credit card lets her buy goods and pay for them at the end of the month. Failure to pay then means bank charges of 2% per month on the amount owing. What interest would Anna have to pay for a month's interest on: **a** £45   **b** £300?

4A The Shady Deal Finance Company lends money at 18% per annum. How much interest does it charge on:
  a £1500 for 12 months   **b** £1000 for 10 months
  c £750 for 5 months   **d** £20 000 for 1 month?

5A Put money in the bank's deposit account, and they'll give you 6% p.a. interest. Ask for a loan and they'll charge you 13% p.a.
  a Why the difference?
  b Calculate the difference between the interest on a deposit of £1000 for 6 months and the charge for a loan for the same period.

6B Just the three–piece suite Viv wants. But how will she pay for it? Take out a loan of £950 for a year at 12% p.a., or use HP?
  a Which way would be cheaper, and by how much?
  b Why might she choose the dearer method?

cash £950, or 10% deposit + 12 monthly payments of £84

COMPOUND INTEREST

Sally Brown put £1000 in the North-South Savings Bank at an annual rate of interest of 10%. Each year the interest is calculated on the previous year's balance, that is 'principal + interest'. This is called *compound interest*.

NORTH-SOUTH SAVINGS BANK				
Date	Code	In	Out	Balance
20.1.85		1000	—	1000
20.1.86	INT.	100	—	1100
20.1.87	INT.	110	—	1210
20.1.88	INT.	121	—	1331

Some Building Societies calculate it every six months; some lending companies calculate it monthly. Usually, banks calculate the interest on pounds only.

═══════════════════ *Exercise 3* ═══════════════════

1A George is calculating how much money he will have in his Investment Account after 3 years. He has deposited £800 at 11% per annum, and will leave it to grow with compound interest.
Copy and complete his calculations.

First year:　　Principal = £800
　　　　　　　Interest　= 11% of £800 = £88
Second year:　Principal = £888
　　　　　　　Interest　= 11% of £888 = ...
Third year:　　Principal = £985·68
　　　　　　　Interest　= . . . .
　　　　　　　Total sum of money in account = £ . . . . .

2A Calculate the compound interest, as in question **1A**, on:
  a £150 for 2 years at 3% per annum
  b £260 for 3 years at 5% per annum
  c £2500 for 3 years at 9% per annum.

A short-cut In question **1A**, the principal each year is multiplied by the factor 1·11. Using the 'constant facility' on your calculator can help.

$$£800 \rightarrow 1·11 \times £800 \quad = £888 \quad \text{(after 1 year)}$$
$$\rightarrow 1·11 \times £888 \quad = £985·68 \quad \text{(after 2 years)}$$
$$\rightarrow 1·11 \times £985·68 = £1094·10 \quad \text{(after 3 years).}$$

3A Use this short-cut to answer question **2A** again.

4A Calculate the compound interest on £3456 for 10 years at 7·5% p.a.

5A How many years would it take for £100 to double itself at 10% per annum compound interest?

6B

Compound your interest with the Handshake Building society

Erica deposits £850 with the Handshake Building Society, which pays 8% interest half–yearly.
a How much will she have altogether at the end of one year?
b Compare this with an *annual* 8% interest rate.

7B A local council takes out a loan of £1 400 000 to pay for a new swimming pool. The annual rate of compound interest is 13%. How much does the council owe at the end of 4 years if none of the loan has been repaid?

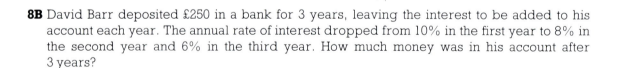

8B David Barr deposited £250 in a bank for 3 years, leaving the interest to be added to his account each year. The annual rate of interest dropped from 10% in the first year to 8% in the second year and 6% in the third year. How much money was in his account after 3 years?

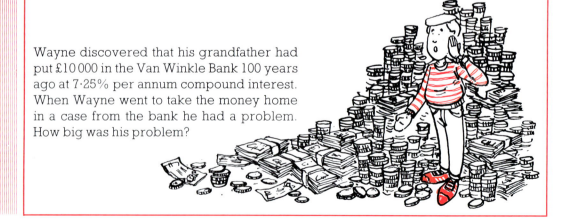

Wayne discovered that his grandfather had put £10 000 in the Van Winkle Bank 100 years ago at 7·25% per annum compound interest. When Wayne went to take the money home in a case from the bank he had a problem. How big was his problem?

ANNUAL PERCENTAGE RATE (APR)

In the advertisement the monthly rate of interest is 2%. But the *annual* rate is more than $12 \times 2\%$, as it is calculated on the **loan + interest each month**. This means that the interest is compounded *monthly*.

=== *Exercise 4* ===

1A Use the 'constant facility' on your calculator to check that a loan of £100 at 2% rate of interest per month becomes £126·82 at the end of 12 months. So the APR is 26·82%.

2A Some credit–card companies charge 3% per month. Calculate the APR.

3A

Is Honest Joe really honest? He should say what the APR is. Can you work it out for him?

4A Calculate the APR for a monthly interest rate of: **a** 1%   **b** 0%.

5B A money-lender charges an APR of 1000%. Calculate the monthly rate of interest, to the nearest whole number, needed to reach this figure. Deceptive, isn't it?

Mr Smith takes out a loan for a year at a monthly rate of 3·5%. At the end of the year he clears his debt by paying £302·21. How much did he borrow?

Investigate the APRs charged by different credit cards, shops, garages and finance companies. The rates are often given in newspaper advertisements. What are the highest and lowest APRs that you can find?

LIFE ASSURANCE

Class discussion

which Kind of life assurance policy Should I have?

Look at these three Kinds of policy

Whole life policy

Pays a fixed amount to your dependants when you die.

Term policy

For a fixed number of years.
Pays a fixed amount to your dependants if you die during the term.
Useful for covering House Mortgages (Chapter **8**).

Endowment policy

For a fixed number of years.
Pays a fixed amount (or more) to your dependants if you die during this time.
Pays you the money if you survive this period, usually with profits.

What are the advantages and disadvantages of each kind of policy?
(See also the tables of costs below.)

The Safe and Secure Insurance Company calculates the premiums you pay from these tables.
Most companies give different rates for men and women.

Monthly premiums for every £1000 assured

WHOLE LIFE (with profits)

Age Now	Non-Smoker	Smoker
25	£1·80	£2·20
26	1·80	2·20
27	1·80	2·40
28	1·80	2·60
29	1·85	2·80
30	1·90	3·00
31	2·00	3·20
32	2·10	3·40
33	2·25	3·60

ENDOWMENT (with profits)

Age Now	10 years		20 years	
	Non-Smoker	Smoker	Non-Smoker	Smoker
18–23	£9·10	£10·60	£4·50	£5·35
24	9·20	10·75	4·65	5·50
25	9·35	10·95	4·80	5·65
26	9·50	11·15	5·00	5·90
27	9·60	11·35	5·20	6·15
28	9·70	11·45	5·40	6·40
29	9·85	11·55	5·50	6·55
30	10·00	11·70	5·65	6·75
31	10·15	11·85	5·85	6·95

SAVING AND SPENDING

1A Why do the premiums in both tables increase as age increases?

2A a Why is there a difference between the premiums for smokers and non–smokers?
b At age 25, the smoker's whole life premium is £2·20. How much older has a non–smoker to be to pay the same or more?

3A Andrea, 21 years old and a non–smoker, works in a library. She takes out an endowment policy worth £7500. From the table, her monthly premium for £1000 assurance is £4·50.

Copy and complete the following calculation of her monthly premium:

Assurance	Premium
£1000 ⟷	£4·50
£7500 ⟷	£4·50 × $\frac{7500}{1000}$ = £ . . . .

Her monthly premium is £ . . . .

4A Calculate the premiums for these people:

Name	Age	Policy	Term	Sum assured
Justin Case (non-smoker)	26	Whole life	Life	£10 000
Fallin Downie (smoker)	22	Endowment	20 years	£6000
Al Dunn (smoker)	18	Endowment	10 years	£5000
I. McClean (non-smoker)	32	Whole life	Life	£25 000

5A Jim Thomson, 31, a keen sportsman, doesn't smoke. He takes out a 20 year endowment policy, with profits, giving £8000 cover.
a Calculate his annual premium.
b How much less would he pay for a whole life policy?

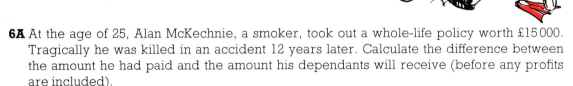

6A At the age of 25, Alan McKechnie, a smoker, took out a whole-life policy worth £15 000. Tragically he was killed in an accident 12 years later. Calculate the difference between the amount he had paid and the amount his dependants will receive (before any profits are included).

7A

Christine, a 26 year old dentist, finds she can save £80 a month. She decides to put this into life assurance. What choices has she? How much cover can she get (to the nearest £100)? What difference does it make whether she smokes or not?

8A At 29, Jean Sharp, a non–smoker, took out a 10 year endowment policy worth £4500. Calculate:
a the amount she paid in premiums over 10 years
b the amount she received from the insurance company at the end of 10 years, assuming that they paid her £1655 profits.

9A Habib's hobby is hang–gliding. He is 28, and doesn't smoke. Safe and Secure will give him whole life cover of £10 000 provided he pays 40% extra for his premiums.
a Why has he to pay 40% extra?
b Calculate his annual premium.

10B Tim, a 25 year old, doesn't smoke. He has £1000 whole life cover. After how many years will he have paid £1000 in premiums?

11B The Safe and Secure Insurance Company realises that inflation will eat into the value of the money they pay out. So they offer a new 10 year endowment policy. For £1000 cover, the first year's premium for a 20 year old non–smoker is £90. Both premiums and cover increase each year by 10% of their previous year's value. Calculate the premium and the cover for:
a the second year   **b** the third year.

12B Sheila Young has a 20 year endowment policy for £5500, plus profits. She is 29 years old, and a non–smoker. After 3 years she decides to stop paying and surrenders her policy, which now has profits of 86 units at £1·875 per unit. She will receive the premiums paid, less 35%, and also the profits, less 20%. What is the 'surrender value' of the policy?

Another type of life or savings policy is 'unit–linked'. Premiums are put into unit trusts or other investment funds, and the cash value of the policy depends on the degree of success of these funds. Investigate the advantages and disadvantages of this type of policy.

CHECK-UP ON **SAVING AND SPENDING**

1 The balance in Ross's bank account is £42·50. What will the balance be after £85 is credited and £120·80 is debited?

2 Pauline has £750 to invest. Should she put it into a Building Society account at 8% p.a. (tax paid), or into a National Savings Investment Account at 12% p.a. (tax to be deducted at 27p in the £)?

3

NS Income Bonds
Interest 12·5% p.a.,
paid monthly

INVESTMENT (£)	MONTHLY INCOME (£)
2000 (minimum)	20·83
5000	52·08
10 000	104·16

Mrs Wright, a retired Bank Manageress, puts £12 000 into Income Bonds.
a (i) Write down her monthly income.   (ii) How much is this in a year?
b Give two reasons for Mrs Wright choosing Income Bonds.

4 Alan is saving for a car. He puts £1500 into a deposit account at 10% p.a. He leaves the interest in so that his money grows with compound interest. How much will he have after:
a 1 year   **b** 2 years   **c** 3 years?

5 An investment of £250 makes £21·25 interest in a year. Calculate the rate of interest.

6 Mike Milligan, a teacher of physical education, is 26 years old and a non-smoker. He can afford £55 a month for a life assurance policy. Use the tables on page 72 to find how much (to the nearest £100) he can insure his life for with:
a a whole life policy
b a 20-year endowment policy.

7 Marion can have a £9000 whole life policy for £1·85 a month, or a £9000 20-year endowment policy for £5·50 a month. What factors should she consider before making a choice?

LOOKING BACK

Reminders

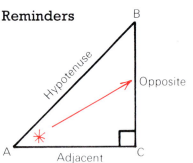

In right-angled $\triangle ABC$,

$$\sin A = \frac{\text{Opposite side}}{\text{Hypotenuse}}$$

$$\cos A = \frac{\text{Adjacent side}}{\text{Hypotenuse}}$$

$$\tan A = \frac{\text{Opposite side}}{\text{Adjacent side}}.$$

Memory aid:

SOH–CAH–TOA

=================== *Exercise 1* ===================

1A Calculate $x°$, to the nearest $0·1°$:

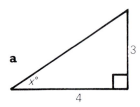

a   **b**   **c**

2A Calculate d, correct to 1 decimal place:

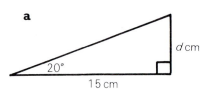

a   **b**   **c**

3A

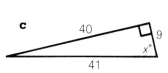

The diameter BD of the circle is 19 cm long.

a What is special about $\triangle ABD$ that allows you to say that $\sin ABD = \frac{6}{19}$?

b Calculate: (i) $\angle ABD$, to the nearest degree
                 (ii) CD, to the nearest centimetre.

4A The trawler 'Seven Seas' sets sail on a course 035° from its home port. It keeps to this course for 25 km. How far:

    a East   **b** North, is it from port, to the nearest kilometre?

5A To reach the observation platform Graeme and Martin have to climb the steps and then the ladder. What height is the platform above the ground, to one tenth of a metre?

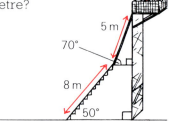

6A

A hole is drilled in the metal sheet at A. Another hole has to be drilled at B. Calculate the coordinates of B, to 1 decimal place.

7A The Sun is 9.3×10^7 miles from the Earth, and the Moon is 2.4×10^5 miles from the Earth. Calculate $\angle$ ESM when $\angle$ SEM = 90°.

8A PQR is a right–angled isosceles triangle.

a Prove that $\sin 45° = \dfrac{1}{\sqrt{2}}$.

b Write down values of: (i) $\cos 45°$ (ii) $\tan 45°$.

9A DEF is an equilateral triangle, and FG is an altitude.
a Each side of $\triangle$DEF is 2 units long. Prove that FG = $\sqrt{3}$ units.
b Copy and complete this table:

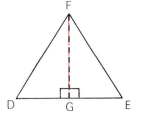

sin 30°	cos 30°	tan 30°	sin 60°	cos 60°	tan 60°
	$\dfrac{\sqrt{3}}{2}$			$\dfrac{1}{2}$	

10B

Calculate the area of this end of the lean–to greenhouse, to the nearest square metre.

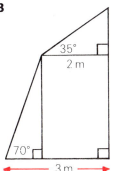

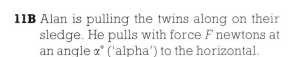

11B Alan is pulling the twins along on their sledge. He pulls with force F newtons at an angle $\alpha°$ ('alpha') to the horizontal.

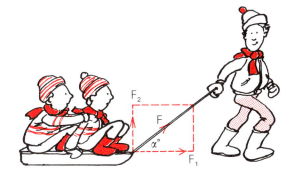

a Prove that the horizontal and vertical 'components', or parts, of the force are $F \cos \alpha°$ newtons and $F \sin \alpha°$ newtons.
b Calculate, to 1 decimal place, the components when $F = 50$ and α is:
(i) 20 (ii) 45 (iii) 70.
c Explain what happens when: (i) $\alpha = 0$ (ii) $\alpha = 90$.

12B

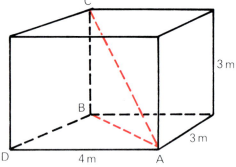

A goods container is in the shape of a cuboid. Calculate:
a the length of:
   (i) AB   (ii) AC, to 1 decimal place
b the size of ∠BAC, to the nearest degree.

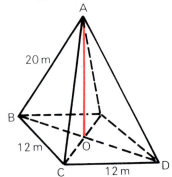

13B The drilling rig has a vertical pipe OA rising from the centre of a square base. Calculate the angle between one of the sloping support struts and the ground (∠ABO), correct to 0·1 degree.

14B

The top of the cooker is horizontal. Calculate the angles, to the nearest degree, between the vertical QP and:
a PR
b PS
c PT.

SINES, COSINES AND TANGENTS OF ALL SIZES OF ANGLES

Coordinates help trigonometry to break free from the shackles of the right–angled triangle.

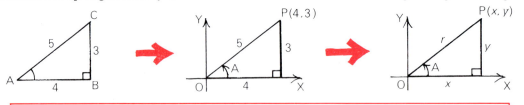

Using coordinates, for any angle A, $\sin A = \dfrac{y}{r}$, $\cos A = \dfrac{x}{r}$, $\tan A = \dfrac{y}{x}$.

Example

a $\sin A = \frac{3}{5}$ **b** $\sin A = \frac{3}{5}$ **c** $\sin A = -\frac{3}{5}$ **d** $\sin A = -\frac{3}{5}$.

Note (i) If OP = *r* rotates *anti-clockwise* from OX through an angle A, the angle is positive.
(ii) If OP = *r* rotates *clockwise* from OX through an angle A, the angle is negative.
(iii) *r* is always positive.

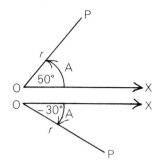

═══════════════ *Exercise 2* ═══════════════

1A Copy the diagrams in the *Example* on page 78, and write down the values of cos A and tan A for each one.

2A Plot each position of P on squared paper, and draw ∠XOP. Write down the 'trig' ratios' for sin XOP, cos XOP and tan XOP, for:
    a P(5, 12)   **b** P(−5, −12)   **c** P(−5, 12)   **d** P(5, −12).

3A a Draw a circle centre O, radius 5 cm, on squared paper.
    b Draw OX horizontally, and ∠XOP = 40°, P on the circle.
    c By measuring suitable lines, calculate, to 2 decimal places, approximations for sin 40°, cos 40° and tan 40°.
    d Check your answers with your calculator.

4A Using the same circle, do question **3A** again for:
    a ∠XOP = 120°   **b** ∠XOP = 200°   **c** ∠XOP = 310°.

5B a Using your answers to questions **3A** and **4A**, copy and complete this table:

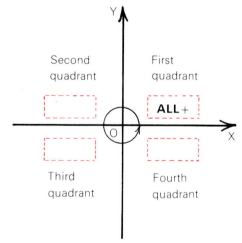

Angle	Quadrant	Is the SIGN + or −?		
		sine	cosine	tangent
40°	First	+		
120°			−	
200°				+
310°				

    b Copy the quadrants diagram, and say which of SINE, COSINE or TANGENT is positive in each quadrant.

6B a Without using your calculator, say which of the following are positive, and which are negative:
    (i) sin 100°   (ii) cos 100°   (iii) tan 100°   (iv) cos 290°   (v) sin 355°   (vi) tan 191°
    b Check your answers with your calculator.

1 Draw diagrams to show the position of OP
when A is:
0°, 90°, 180°, 270°, 360°.
In each case mark the coordinates of P,
and investigate the values of the sines,
cosines and tangents of these angles.
What happens for angles greater than 360°?

2 Use your answers to **1** and to questions **8A** and **9A** on page 77 to copy and
complete this table:

A	0°	30°	45°	60°	90°	180°	270°	360°
sin A								
cos A								
tan A					—		—	

REPEATING PATTERNS

Repeating patterns such as shape, time and distance are all around us.

The days of the week Period 7 days

$\frac{1}{7} = 0.142857, 142857...$ *Period 6 figures*

car mileometers... [9][9][9][9][9] *Period 100,000 miles*

=== *Exercise 3* ===

1A Describe any repeating
patterns you can see in
the picture.

2A Use your calculator to
write each fraction as a
decimal. Then write
down the *period* of the
repeating pattern of
numbers.

  a $\frac{2}{11}$ **b** $\frac{41}{333}$ **c** $\frac{4}{9}$

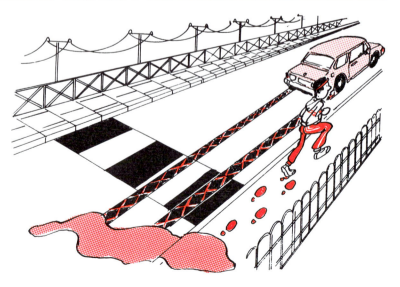

3A 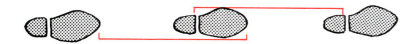 Measure the *period* of this set of footprints in the snow.

4A The ferry goes back and forward across the river all day long.

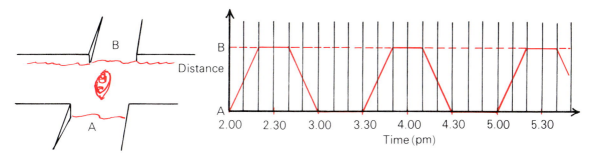

a How long does it spend: (i) at each side   (ii) crossing?
b What is the period of its operation?

5A

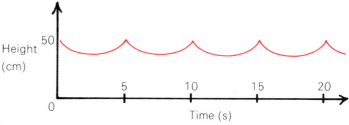

Swing high, swing low . . . . What is the period of the swing, in seconds, from one side to the other and back again?

6A Write down other examples of repeating patterns of shape, time, distance, movement, etc.

7B a Copy and complete the table:

Number	0	1	2		9	10	11
Remainder when divided by 4	0	1	2				3

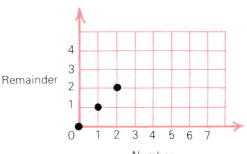

b Copy and complete the graph.
c What is the period of the graph?

THE TRIGONOMETRIC FUNCTIONS AND THEIR GRAPHS

For each angle A, there is only one value of sin A between −1 and 1 inclusive. So there is a function which maps the set of all angles to the set of all numbers from −1 to 1. This function is called the **sine function**.

The **cosine function** is described in a similar way.

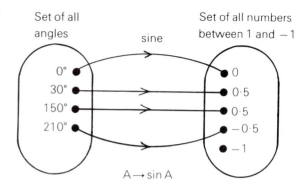

$A \rightarrow \sin A$

=========== Exercise 4 ===========

1A The graph of the sine function

a On 2 mm squared paper, draw a circle with radius 2 cm. Call the radius 1 unit long. Draw $\angle XOP_1 = 30°$, $\angle XOP_2 = 60°$, $\angle XOP_3 = 90°$, and so on, as OP rotates round O through 360° from OX.

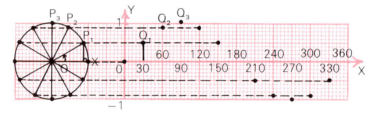

$\mathrm{Sin}\, XOP_1 = \dfrac{y_1}{r} = \dfrac{y_1}{1} = y$, so the y–coordinates of P_1, P_2, P_3, . . . give the values of sin 30°, sin 60°, sin 90°, ...

b Mark 30, 60, 90, . . . on the x-axis and draw horizontal lines through P_1, P_2, P_3, . . . to give the values of sin 30°, sin 60°, sin 90°, . . . at Q_1, Q_2, Q_3, ...

c Join O, Q_1, Q_2, Q_3, . . . with a smooth curve. *This is part of the graph of the sine function.*

d Think of OP rotating about O from OX for a second time, through 390°, 420°, . . . , to produce another *cycle* of the sine curve. To see this, trace the 0°–360° curve and slide your tracing paper to the right, parallel to the x-axis. Doing this repeatedly gives a smooth curve called the **sine curve**, or **sine wave**.

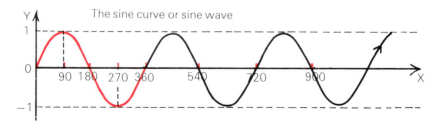

The sine curve or sine wave

Notice that the sine curve repeats itself for every interval of 360°. So if 360°, or multiples of 360° are added to A (or subtracted from A), the value of sin A remains the same. For this reason, the sine function is a **periodic function** with **period** 360°.

e Check from the graph that for all angles $x°$, $-1 \leqslant \sin x° \leqslant 1$.

2A The graph of the cosine function

a Copy and complete this table, using your calculator.

x	0	30	60	90		300	330	360
$\cos x°$	1	0·87	0·50			0·50	0·87	1

b Plot the points, and draw the cosine curve.

c Use your calculator to check that you will get the same values for $\cos x°$ by going from 360° to 720° in 30° steps.

d So **the period of the cosine function is 360° also.** Trace the cosine curve you have drawn and slide it to the right to check this.

e Try to fit the tracing over the sine graph. Do you find that the graphs are similar and 'out of phase' by 90°?

f Write down the maximum and minimum values of the cosine function and the smallest corresponding values of x.

3A a What do you think the graph of $y = 2 \sin x°$ looks like?

b Copy and complete the table.

c Sketch the graph on plain paper, marking your scales.

d Sketch $y = 2 \cos x°$ on plain paper also.

x	0	90	180	270	
$\sin x°$	0	1			
$2 \sin x°$	0	2			

4A a What will the graph of $y = \sin (x+90)°$ look like?

b Copy and complete the table.

c Check the points on the graph of $y = \cos x°$. What do you find? Can you write $\sin (x+90)°$ in a simpler form?

x	0	90	180	270	360
$x+90$					
$\sin (x+90)°$					

d Try the same for the graph of $y = \cos (x+90)°$, compared to $y = \sin x°$.

5B Using a table of values with $x = 0, 30, 60, 90, \ldots, 360$, draw the graphs of:

a $y = \sin x°$, $y = \sin 2x°$ and $y = \sin 3x°$ with the same axes.

b $y = \cos x°$, $y = \cos 2x°$ and $y = \cos 3x°$ on another diagram.

6B a It is claimed that, 'The function S defined by $S(x) = \sin 2x°$ has the same maximum and minimum values as the function $s(x) = \sin x°$, but it has half the period'. True or false?

b The same statement is made about the function C for which $C(x) = \cos 2x°$ and c for which $c(x) = \cos x°$. Is this statement true or false?

The graph of $y = \sin nx°$ repeats when nx is increased by 360, that is, when x is increased by $\dfrac{360}{n}$.

So the period of the function whose values are given by $y = \sin nx°$ is $\dfrac{360°}{n}$.

Similarly, $\dfrac{360°}{n}$ is the period of the function for which $y = \cos nx°$.

In the equations: $y = a \sin nx°$ and $y = a \cos nx°$
  (i) a determines the maximum and minimum values
 (ii) n determines the period $\dfrac{360°}{n}$.

7B Write down the maximum and minimum values, and periods, of the functions given by:
  a $y = 4 \sin x°$ **b** $y = \cos 2x°$ **c** $y = 5 \sin x°$ **d** $y = 3 \cos x°$
  e $y = 2 \cos 3x°$ **f** $y = \frac{1}{2} \sin 3x°$ **g** $y = \sin (x+80)°$ **h** $y = 10 \cos (x-100)°$.

8B Can you work out the equations of these graphs?

a   **b**

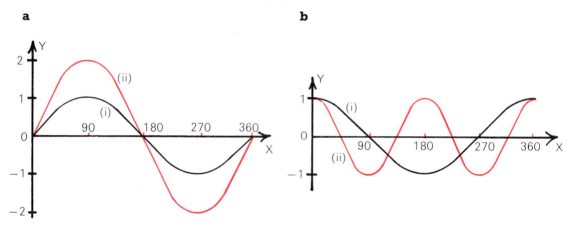

9B The graph of the tangent function
Copy and complete this table, using your calculator.

x	0	30	60	75	105	120	150	180	210	240	255	285	300	330	360
$\tan x°$	0	0·58													

  a Plot the sets of points between the double lines in the table and join each set of points with a smooth curve.
  b Key in tangents of angles near 90°, e.g. 89°, 89·9°, 89·999999°, 90·000001°. Draw vertical lines through 90° and 270°. As x increases from zero, the graph gets closer and closer to these lines which are called *asymptotes*.

The graph of $y = \tan x°$ consists of an endless number of branches. These occur at intervals of 180°, so the period of the tangent function is 180°. Tan $x°$ has no value at $x = 90, 270, \ldots$ so the function is undefined at these points.

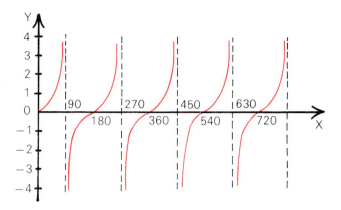

INVESTIGATION

1 The sine wave is useful in physics. Try to find out where, and see it on an oscilloscope. Perhaps you can produce it on a computer screen, and vary its shape and size.

2 You need a ruler, protractor and 2 mm squared paper.
  a Draw axes, with a scale of 4 mm to 10° on the x–axis.

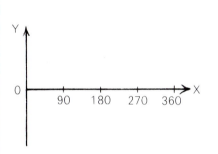

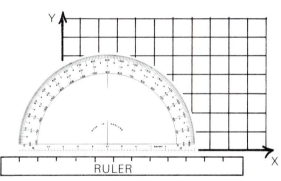

RULER

  b Place the ruler and protractor as shown.
  c Slide the protractor to the right, and mark the point where the end of each radius meets the corresponding vertical. In the diagram the 30° radius meets the 30 vertical.
  d At 180, turn the protractor upside down, and continue to 360.
  e Join the points with a smooth curve.
  f Investigate why it is the sine curve.

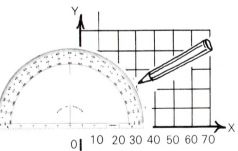

  a On the same diagram sketch the graphs of $y = \sin x°$ and $y = \cos x°$ for $0 \leqslant x \leqslant 360$.
  b List x for which: (i) $\sin x° = \cos x°$ (ii) $\sin x° > \cos x°$ (iii) $\sin x° < \cos x°$.
  c For what values of x is $\sin x° + \cos x°$:   (i) a maximum   (ii) a minimum?

6

A surprising discovery

1A Pat had turned off . . . . The afternoon was hot, and Mr Grumble droned on. She idly keyed in sin 40° and cos 50° on her calculator, then sin 30° and cos 60°, sin 10° and cos 80°. Suddenly she looked interested. Why?

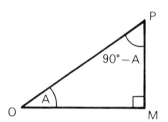

   a Key in these pairs on your calculator.
   b What do you find?
   c How are the angles related?
   d Try some more of your own.

2A Mr Grumble agreed that Pat had found a special connection between sines and cosines of angles. He asked her to *prove* it. Copy and complete her proof:

The two angles add up to . . . .°

$$\sin A = \frac{MP}{\ldots}$$

$$\cos(90° - A) = \frac{\ldots}{\ldots} = \sin \ldots$$

The sine of an angle = the **co**sine of its . . . .

> $$\sin(90° - A) = \cos A, \text{ and } \cos(90° - A) = \sin A$$

3A **a** Write each of these in the form cos $x°$:　(i) sin 45°　(ii) sin 88°　(iii) sin 8·5°.
   b Write each of these in the form sin $x°$:　(i) cos 77°　(ii) cos 1°　(iii) cos 29·9°.

4A Pat's cosine key refused to work. Help her to calculate x without it:

a

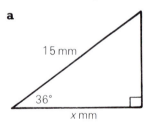

b

c

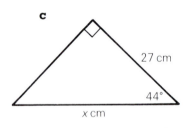

Another discovery

Pat had got a new calculator and was keen to try it out.

THESE were complementary angles. How about supplementary angles?

1A Pat started exploring . . . sin 20°, sin 160°; cos 20°, cos 160°; tan 20°, tan 160°. Unexpected results! Find out for yourself.

2A a Copy and complete the table, using your calculator. Make entries correct to 3 decimal places.

	20°	160°	50°	130°	100°	80°	123°	57°
sine								
cosine								
tangent								

b What do you notice about:
(i) the pairs of angles   (ii) the patterns in the table?

3A

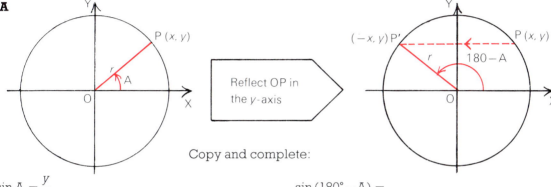

Copy and complete:

$$\sin A = \frac{y}{r}$$

$$\cos A = \dots$$

$$\tan A = \dots$$

$$\sin(180° - A) = \dots$$

$$\cos(180° - A) = \dots$$

$$\tan(180° - A) = \dots$$

These formulae connect *supplementary angles*.

$$\sin(180° - A) = \sin A, \cos(180° - A) = -\cos A \text{ and } \tan(180° - A) = -\tan A$$

4A Copy and complete, putting + or − in each box:

a $\sin 140° = \square \sin 40° = 0\cdot\dots$ **b** $\cos 140° = \square \cos 40° = \dots$

c $\tan 140° = \square \tan 40° = \dots$ **d** $\cos 110° = \square \cos 70° = \dots$

5A Using the method in question **4A**, calculate the values of:

a $\sin 150°$ **b** $\cos 118°$ **c** $\tan 175°$ **d** $\sin 106°$
e $\cos 99°$ **f** $\tan 123°$ **g** $\sin 147\cdot5°$ **h** $\cos 158\cdot3°$.

6A Part of the sine curve has been drawn.
a Describe the axis of symmetry.
b Why does the graph show that $\sin 150° = \sin 30°$?
c $\sin 120° = \sin \dots$ Which angle?

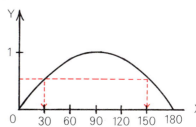

7A

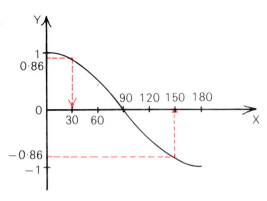

Part of the cosine curve.
a Where is the centre of symmetry?
b Why does the graph show that
$\cos 150° = -\cos 30°$?
c $\cos 120° = -\cos \dots$ Which angle?

TRIGONOMETRIC EQUATIONS

Example
Solve these equations, to the nearest degree, for $0 \leqslant x \leqslant 360$:
a (i) $\sin x° = 0.6$ (ii) $\sin x° = -0.6$ **b** (i) $3 \cos x° - 1 = 0$ (ii) $3 \cos x° + 1 = 0$.

a

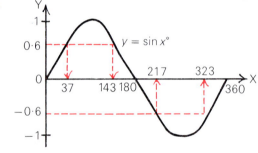

(i) Using a calculator,
$\sin x° = 0.6 \Rightarrow x = 37$, to the nearest whole number.
From the symmetry of the $0-180$ part of the sine curve, $x = 180 - 37 = 143$ also.
So $x = 37$ and 143.
(ii) From the symmetry of the curve about 180, $\sin x° = -0.6 \Rightarrow x = 180 + 37 = 217$ and $360 - 37 = 323$.

b

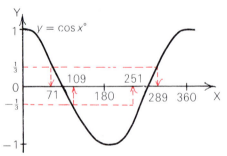

(i) $3 \cos x° - 1 = 0$, so $\cos x° = \frac{1}{3}$
$x = 71$, to the nearest whole number.
From the symmetry of the cosine curve,
$x = 360 - 71 = 289$ also.
So $x = 71$ and 289.
(ii) $3 \cos x° + 1 = 0$, so $\cos x° = -\frac{1}{3}$.
From the symmetry of the curve,
$x = 180 - 71 = 109$
and $180 + 71 = 251$.

Note The 'all-sin-tan-cos' diagram can be useful in solving equations like the ones above, as shown in examples **a** and **b** here.

sin +	all +
tan +	cos +

a

(i)

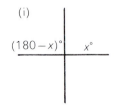

(ii)

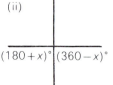

b

(i)

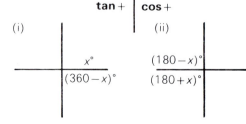

(ii)

Exercise 7

Solve these equations to the nearest degree, for $0 \leqslant x \leqslant 360$. Then check your answers with a calculator.

1A a $\sin x° = 0.5$ **b** $\sin x° = -0.5$ **2A a** $\cos x° = 0.7$ **b** $\cos x° = -0.7$

3A a $2\sin x° = 0.4$ **b** $5\sin x° = -0.8$ **4A a** $4\cos x° = 3$ **b** $7\cos x° = -4$

5A $4\sin x° + 1 = 0$ **6A** $3\cos x° - 2 = 0$ **7B** $5\sin x° - 1 = \sin x°$

8B $5\cos x° + 2 = \cos x°$ **9B** $\sin^2 x° = 0.25$ **10B** $\cos^2 x° = 0.64$

11B $\tan x° = 1$ **12B** $\tan x° = -2$ **13B** $\tan^2 x° = 100$.

TWO USEFUL FORMULAE

1 Connecting $\sin A$, $\cos A$ and $\tan A$

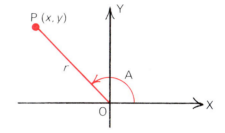

$$\frac{\sin A}{\cos A} = \frac{\frac{y}{r}}{\frac{x}{r}} = \frac{y}{r} \times \frac{r}{x} = \frac{y}{x} = \tan A.$$

$$\boxed{\tan A = \frac{\sin A}{\cos A}, \cos A \neq 0}$$

Exercise 8

1A a Write down fractions for $\sin A$, $\cos A$ and $\tan A$.

    b Check that $\dfrac{\sin A}{\cos A} = \tan A$.

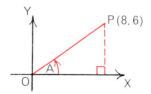

2A Repeat question **1A** for angle A an obtuse angle, with P the point $(-5, 12)$.

3A To solve the equation $\sin x° = \cos x°$, where $0 \leqslant x \leqslant 90$, copy and complete:

$\sin x° = \cos x°$

$\dfrac{\sin x°}{\cos x°} = \dfrac{\cos x°}{\cos x°}$ (Divide each side by $\cos x°$.)

So $\tan x° = \ldots$

$x = \ldots.$

4B a For each equation, calculate $\tan x°$. *(Hint:* Divide every term by $\cos x°$.)

    b Then find x in each, to 1 decimal place, where $0 \leqslant x \leqslant 90$.

      (i) $\sin x° = 2\cos x°$ (ii) $2\sin x° - 6\cos x° = 0$ (iii) $4\sin x° + 3\cos x° = 2\sin x° + 4\cos x°$.

5B Use the graph of $y = \tan x°$ on page 85 to calculate another value for x in each equation in question **4B** for $0 \leqslant x \leqslant 360$.

6B a Simplify: $\sin A \times \cos A \times \tan A$.   **b** Prove that: $\dfrac{\cos A}{\sin A} \times \tan A = 1$.

2 Connecting sin A and cos A

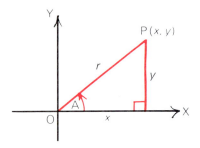

$x^2 + y^2 = r^2$ (Pythagoras' Theorem)

$\dfrac{x^2}{r^2} + \dfrac{y^2}{r^2} = 1$ (Divide by r^2)

$\left(\dfrac{x}{r}\right)^2 + \left(\dfrac{y}{r}\right)^2 = 1$

So $(\sin A)^2 + (\cos A)^2 = 1$, which is usually written:

$$\sin^2 A + \cos^2 A = 1$$

=== *Exercise 9* ===

1A a Prove that $\sin A = \sqrt{(1 - \cos^2 A)}$, $0° \leqslant A \leqslant 90°$.
  b Use this formula to calculate sin A, correct to 3 decimal places, when $\cos A = 0 \cdot 700$.

2A Check with your calculator that:
  a $\sin^2 55° + \cos^2 55° = 1$ **b** $\sin^2 1 \cdot 5° + \cos^2 1 \cdot 5° = 1$.

3A a Prove that $\cos A = \sqrt{(1 - \sin^2 A)}$, $0° \leqslant A \leqslant 90°$.
  b Use your formula to calculate cos A, correct to 3 decimal places when sin A has the value:
    (i) $0 \cdot 4$ (ii) $0 \cdot 567$.

4A $\sin^2 A = 0 \cdot 64$. Calculate: **a** $\cos^2 A$ **b** $\tan^2 A$.

5B Prove that:
  a $3\sin^2 A + 3\cos^2 A = 3$ **b** $\sin x \cos^2 y + \sin x \sin^2 y = \sin x$
  c $(\sin A + \cos A)^2 = 1 + 2\sin A \cos A$ **d** $(\sin P + \cos P)^2 + (\sin P - \cos P)^2 = 2$
  e $(x\cos A + y\sin A)^2 + (x\sin A - y\cos A)^2$ is independent of angle A.
  f The point $(2\cos A, 2\sin A)$ lies on the circle with equation $x^2 + y^2 = 4$.
  g $\cos^2 A - \sin^2 A = 1 - 2\sin^2 A = 2\cos^2 A - 1$.

CHECK-UP ON **TRIGONOMETRY**

1 Calculate x, correct to 1 decimal place:

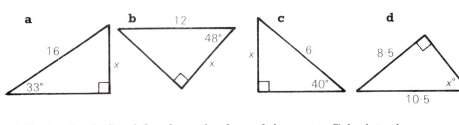

a   **b** 12   **c**   **d**

16   48°   6   8·5

x x x $x°$

33°   40°   10·5

2 Each wire is fixed 8 m from the foot of the mast. Calculate the distance between the points where the wires are fixed to the mast, to the nearest tenth of a metre.

72°   66°
8 m   8 m

3

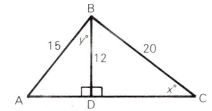

a Use Pythagoras' Theorem to calculate the length of AD.

b Calculate: (i) $\sin x°$ (ii) $\sin y°$.

c Prove that:
(i) $x = y$ (ii) $\angle ABC = 90°$ (in two ways).

4 Plot these points on squared paper, and calculate $\sin$ XOP, $\cos$ XOP and $\tan$ XOP, in their simplest form:

   a P(8, 6)   **b** P(8, −6)   **c** P(−8, 6)   **d** P(−8, −6).

5 The graph shows the brightness of a double star. One star circles another, causing regular eclipses.

   a What is the period of the graph?

   b Explain the two 'dips' in the graph.

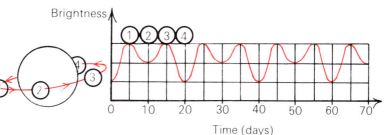

6 On plain paper sketch the graphs of:

   a $y = \sin x°$ **b** $y = \cos x°$ **c** $y = \sin(x + 180)°$ **d** $y = \cos 4x°$.

7

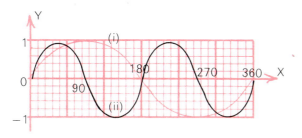

Write down:

   a the equations

   b the periods,
   of these two graphs.

8 Copy and complete:

   a $\sin(90° - A) = \ldots$ **b** $\cos(90° - A) = \ldots$ **c** $\sin 66° = \cos \ldots$ **d** $\cos 15° = \sin \ldots$

9 Solve each equation for $0 \leqslant x \leqslant 180$, correct to 1 decimal place if necessary:

   a $\sin x° = 0·5$ **b** $\sin x° = 0·75$ **c** $\cos x° = 0·5$ **d** $\sin x° = 4\cos x°$.

10 Prove that $(\sin A - \cos A)^2 = 1 - 2\sin A \cos A$.

11 As it moves up and down in the cylinder, the piston moves the connecting rod PA, so turning the crank-shaft OA.

   OA = 6 cm, and is horizontal at the start.

   a In the first three turns of the crank-shaft, how often is A:
   (i) 3 cm to the left of the centre line   (ii) 4 cm to the right
   (iii) at the 'top dead centre', B?

   b Calculate, to 0·1 degree, the angle turned through by OA to the positions in **a** (i), (ii) and (iii).

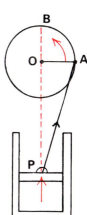

EXPERIMENT AND THEORY

In science, medicine, economics, business and many kinds of research it is often necessary to find whether or not two quantities are related. A theory is put forward, and then tested by experiment. Mathematical models in the form of graphs and equations are used to test the relationship between the quantities. Here are three famous examples from science.

In the nineteenth century, George Ohm, a German scientist, investigated the relationship between the *current* and *voltage* in an electrical circuit.

In the seventeenth century, Isaac Newton studied the connection between the *force* on a body and the *acceleration* it produced.

Another seventeenth–century scientist, Hooke, looked at the relationship between the *tension* in a spring and its *extension*.

The work of all three scientists is summed up in this 'mathematical model'. The quantities they studied are connected by straight line graphs which pass through the origin.

From Chapters **1** and **4** you'll remember the equation of this line, **$y = ax$** —*x and y* 'behave' in the same way—double one, you double the other; halve one, you halve the other; and so on.

> *y* is *directly proportional* to *x*, or *y varies directly* as *x*.
> $y \propto x$ ('*y* varies as *x*'), or $y = kx$, where *k* is the variation constant.

1A *The results of an experiment by Ohm*

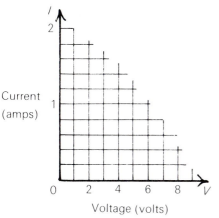

Current
(amps)

Voltage (volts)

Voltage (*V* volts)	0	2	4	6	8	10
Current (*I* amps)	0	0·7	1·4	2·2	2·8	3·5

a Draw the graph of *I* against *V*, using the scales and axes shown.
b Allowing for experimental error, check that the points lie on a straight line through the origin.
Compare $y = ax$.
Here $I = kV$, k a constant.

Current is proportional to voltage, or current varies as voltage.

2A *An experiment like Newton's*

Acceleration (*a* m/s²)	0	5	10	15	20	25	30
Force (*F* newtons)	0	0·4	0·8	1·1	1·6	2·1	2·4

a Draw the graph of *F* against *a*.
b Check that the points lie close to a straight line through O.
c Compare $y = ax$. Copy and complete:
　(i) $F = \ldots\ldots$
　(ii) Force is proportional to . . . . . ., or force . . . .

Force
(N)

Acceleration (m/s²)

3A *Hooke's experiment*

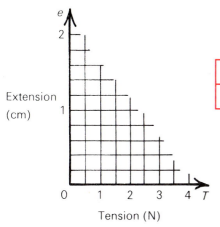

Extension
(cm)

Tension (N)

Tension (*T* newtons)	0	1	2	3	4	5	6
Extension (*e* cm)	0	0·6	1	1·4	2	2·6	3

a Draw the graph of *e* against *T*.
b Check that the points lie close to a straight line through O.
c Write down the relation between *e* and *T* which closely fits the data:
　(i) as an equation　(ii) in words.

Can you think of other pairs of quantities that are related to each other like this?

DIRECT PROPORTION (OR VARIATION)

There are many other areas where one quantity is directly proportional to another. Here is an example.

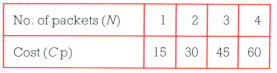

No. of packets (N)	1	2	3	4
Cost (C p)	15	30	45	60

The points lie on a straight line through O.
So $C = kN$.
The cost is directly proportional to the number of packets of crisps.
Look at these ratios from the table:

$$\frac{15}{1} = 15, \frac{30}{2} = 15, \frac{45}{3} = 15, \ldots$$

Always $\dfrac{C}{N} = 15$, so $C = 15N$.

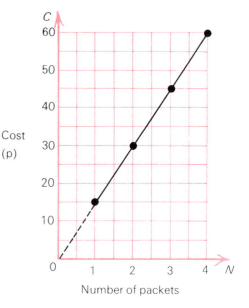

Cost (p)

Number of packets

Direct proportion, or direct variation, can be checked by:
(i) Drawing a graph   (ii) Calculating ratios.

Exercise 2

1A Is the cost of these books proportional to the number purchased?

Number of books (N)	5	10	15	20	25
Cost (£C)	15	30	45	60	75

a Check by calculating values of the ratio $\dfrac{C}{N}$.

b Copy and complete: $C = \ldots N$.

2A Do these earnings vary as the number of hours worked?

Hours (H)	7	14	28	35	40
Earnings (£E)	35	70	140	175	200

a Check by drawing a graph of E against H.

Compare 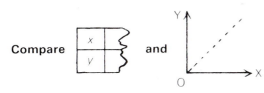 and

So take H on the horizontal axis.

b Copy and complete $E = \ldots H$. (**Compare** $y = ax$.)

3A In the crisps *Example* before Exercise **2**, how much would a packet of crisps be if $k = 1$, and $C = N$? Can you think of two possible answers?

4A

Principal (£P)	0	100	200	300	400
Interest (£I)		10			

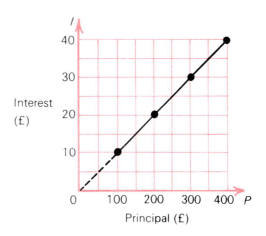

a Use the graph to copy and complete the bank's interest table.
b Why is the interest directly proportional to the principal?
c Copy and complete $I = \ldots P$.

5A Which of these are examples of direct variation?
  a 1 cassette costs £3, 2 cassettes cost £5.
  b 2 bus fares cost 80p, 3 fares for the same journey cost 120p.
  c 1 raffle ticket costs 10p, a book of 10 tickets costs 50p.
  d 3 periods last 165 minutes, 5 periods last 275 minutes.

6A Which of these graphs suggest the model $y = ax$? Give reasons for your answers.

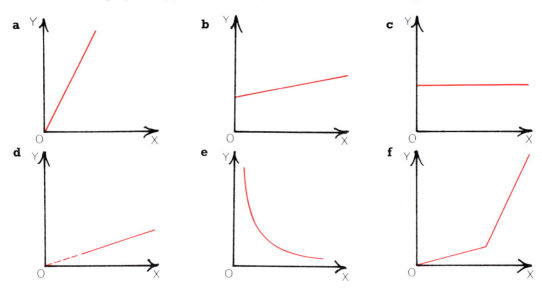

a

b

c

d

e

f

7A Instead of the usual Christmas tree price list, Jim Allan of the Forestry Commission put this graph in the sales shed.

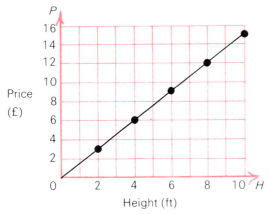

Price (£)

Height (ft)

a Jim claims that his price (£P) is proportional to the height (H feet) of a tree. Do you agree? Give a reason.

b Find a formula for P.

c The tree outside the Town Hall cost £50. Estimate its height.

8A An electric amplifier multiplies the input voltage v to produce a greater output voltage V, i.e. $V = kv$.

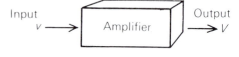

Input v → Amplifier → V Output

An experiment provides this data:

Input voltage (v volts)	1	2	3	4	5	6
Output voltage (V volts)	5	10	15	20	25	30

a By graphing V against v, check that the experiment confirms that $V = kv$.

b What is the value of the amplification factor k?

c What input produces an output of 12 volts?

9B Does the angle turned through by the minute hand of a clock vary directly as the time passed?

a Copy and complete:

Time passed (T minutes)	15	30	45	60
Angle turned ($A°$)				

b Answer the question by: (i) a graph   (ii) calculating ratios.

c Find a formula for A in terms of T.

10B Nick and Harry read that electrical power (W watts) is directly proportional to the voltage (V volts) in the circuit. They set up an experiment and recorded these results:

V	2	3	4	6	10	12	18
W	0·7	0·8	1·1	1·9	6	3·4	5·4

a Plot W against V. Draw the straight line that best fits the data, ignoring the obvious error in their readings. (Which is the error?)

b Do the results agree with what they read? Give a reason.

c Find a formula for W in terms of V.

> Isaac Newton was one of the greatest mathematicians (if not the greatest) of all time. Find out about his life and his discoveries—made through observation, experiment and calculation.

MAKING AND USING PROPORTION MODELS

Example

The 'Silver Star' leaves for the fishing grounds at 02 00, and by 02 50 she has sailed 15 km. Her skipper estimates that the distance from port (D km) is proportional to the sailing time (T minutes).

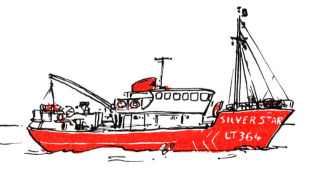

a Find the formula connecting T and D.

b Use the formula to estimate how long the boat will take to reach the fishing grounds, 84 km distant.

D is proportional to T, so $D = kT$.

a When $T = 50$, $D = 15$, so $15 = 50k$

$$k = \frac{15}{50} = 0{\cdot}3$$

and $\boldsymbol{D = 0{\cdot}3T}$.

b When $D = 84$, $84 = 0{\cdot}3T$

$$T = \frac{84}{0{\cdot}3} = 280.$$

The estimated time to reach the fishing grounds is 4 hours 40 minutes.

Exercise 3

1A W is proportional to I. When $W = 1000$, $I = 4$.
  a Find the formula connecting W and I.
  b Calculate W when $I = 3$.

2A The time (T minutes) to serve school lunches varies as the number (N) of pupils taking lunch. When $T = 40$, $N = 200$.
  a Find the formula connecting T and N.
  b Calculate T when $N = 260$.

3A The cost of plate glass for shop windows (£C) varies directly as its area (A m²). The cost of 2 m² is £12.
  a Find the formula connecting C and A.
  b Calculate the cost of 5 m² of glass.

4A Any electrician will tell you that for copper wire of given cross–section the electrical resistance (R ohms) is proportional to the length of the wire (L metres). For example, 100 m of a certain wire has a resistance of 10 ohms.
  a Find the formula connecting R and L.
  b Calculate the resistance of 88 m of wire.

5A Jennifer Lawson travels for a fabric firm. She notices that the cost of petrol she uses (£P) is proportional to the distance (D km) she travels. On Monday 12th October she clocked up 280 km at a cost of £11·20.
  a Construct the formula connecting P and D.
  b Calculate the cost of petrol for a journey of 450 km.

6A In silver plating, the numbers of grams (n) of silver deposited is proportional to the number of minutes (m) the current flows. 2 g of silver are deposited in 45 minutes.
  a Find the formula connecting n and m.
  b Calculate the time required for a deposit of 5 g.

7A The manageress of the local supermarket guesses that the weekly takings (£*T*) vary directly as the number of customers (*N*). During one week, 550 customers spent £4950. Assuming her 'model' is a good one:

a Construct the formula connecting *T* and *N*.

b Estimate the takings for 720 customers.

c How many customers were there in a week when £11 700 was spent?

Eastern Gas Board charges 8p per unit used. Western Gas Board has a standing charge of £10, plus 6d per unit used. Make a table of values for accounts from each Board, and draw two graphs. Which Board has charges that are directly proportional to the number of units used? Give reasons for your answer. Discuss which Board would send the lower bills, according to the number of units likely to be used.

Set up an experiment involving a spring and weights, as in the *Example* on page 93. Make your own table and graph. Write about the results.

DIRECT PROPORTION IN DISGUISE

Exercise 4

1A A new car which is being tested starts from rest and accelerates steadily. The table shows distances, *s* metres, covered in *t* seconds. Find the relation between *s* and *t*.

a By calculating ratios $\frac{2}{1}, \frac{8}{2}, \ldots$ find out if *s* varies as *t*.

b Check by drawing a graph of **s** against **t**. (As shown.)

t	0	1	2	3	4	5
s	0	2	8	18	32	50

The graph looks like a parabola $y = ax^2$, seen in Chapter 1. This suggests $s = kt^2$.

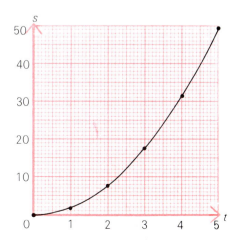

PROPORTION IN PRACTICE

Table of values of **s** against **t²**.

t^2	0	1	4	9	16	25
s	0	2	8	18	32	50

c By calculating ratios $\frac{2}{1}, \frac{8}{4}, \ldots$ find out if s varies as t^2.

d Check by drawing a graph of s against t^2. (As shown.)
The graph is a straight line through O.
So $s = kt^2$.

When $s = 8$, $t^2 = 4$, so $8 = 4k$, and $k = 2$.
The equation is $s = 2t^2$.

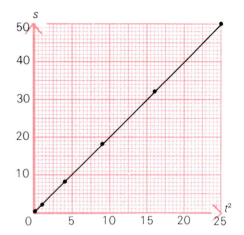

2A Circular cylinders of radii r cm, and fixed height, have volumes V cm³ given by these tables:

r	0	1	2	3	4
V	0	3	12	27	48

r^2	0	1	4	9	16
V	0	3	12	27	48

a Calculate the values of $\dfrac{V}{r}$ and $\dfrac{V}{r^2}$ $(r \neq 0)$. What do you conclude?

b Draw the graphs of V against r, and V against r^2, using the same axes and scales as in question **1A**.

c Show that $V = kr^2$, and calculate k.

3A Is there a connection between the length of a pendulum and the time of its swing?
Alison and Tina set up an experiment.

Length (L cm)	0	9	16	25	36
Time of swing (T s)	0	0·6	0·8	1·0	1·2

'The ratios are not equal' said Alison, 'so T is not proportional to L. What about T against L^2?' Tina shook her head 'Looks more like T against $\sqrt{L}$'.

a Make a new table, with values of $\sqrt{L}$ and T.

b Draw the graph of T against $\sqrt{L}$. ($\sqrt{L}$ on the horizontal axis.)

c Find the formula connecting T and $\sqrt{L}$, and calculate T for $L = 100$.

4A A stone falls D metres in T seconds, and it is known that D varies directly as T^2. After 3 seconds the stone has fallen 45 m.
  a Find the formula for D in terms of T.
  b How far will the stone fall in 6 seconds?
  c How long will it take to fall 125 m?

5A

a

N	2	3	4	5	6
M	12				

b

Q	0	1	4	9	16
P			1		

$M \propto N^2$. Find the formula connecting M and N, and then copy and complete the table.

$P \propto \sqrt{Q}$. Find the formula connecting P and Q, and then copy and complete the table.

6B In the engineering laboratory Tessa and Peter test different materials. They measure the sag (S mm) for different lengths (L m).

L	0	1	2	3	4
S	0	0·5	4	13·5	32

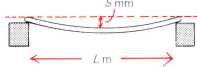

  a Draw the graphs of: (i) S against L (ii) S against L^2 (ii) S against L^3.
  b Which gives a straight line through O? Find the formula connecting S and L.
  c Calculate S when $L = 5$.

7B In each of the following, $y \propto x^n$, where n is a natural number. Find the formula for y in terms of x in each. (Try $n = 1, 2, \ldots$).

a

x	1	2	3	4	5
y	3	12	27	48	75

b

x	0	1	2	3	4
y	0	0·5	1	1·5	2

8B

Too fast round the curve and the train will leave the rails. But how fast? A working model is:
'Safe speed (V m/s) is proportional to the square root of the radius (R m) of the curve.'

  a Construct the formula connecting V and R, given that for a radius of 400 m the safe speed is 40 m/s.
  b Calculate the safe speed for a radius of 80 m (to the nearest m/s).

1 Jacobini, the Swiss diamond cutter, knows that the value of a diamond is directly proportional to the square of its weight. He has to cut a diamond weighing 6 g into two parts in the ratio 2:1. Calculate the percentage gain or loss in value of the diamond.

2 A famous scientist, Kepler, studied the motion of the planets in their orbits round the sun. One of Kepler's 'Laws' is that the square of the time a planet takes to circuit the sun varies directly as the cube of its distance from the sun.

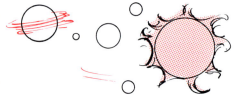

   a The Earth takes 1 year, at a distance of 1·5 million km. Calculate Mars' 'year', at a distance of 2·25 million km from the sun.

   b If the Earth's distance increased by 1%, approximately how many more days would there be in our year?

Set up an experiment involving a pendulum (for example, a variable length of string and a weight), and see whether you agree with Alison and Tina's conclusions on page 100.

INVERSE PROPORTION (OR VARIATION)

The sponsors of Bluebell Rovers' 'Spot the Ball' competition are considering how to divide up the £1200 prize-money.

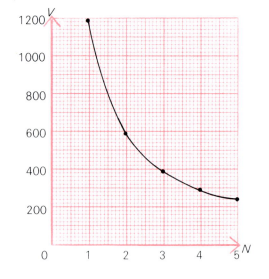

No. of prizes (N)	1	2	3	4	5
Value of each (£V)	1200	600	400	300	240

Is V proportional to N? How can you tell?

The graph looks like a hyperbola $y = \dfrac{a}{x}$,

seen in Chapter 1.

This suggests $V = \dfrac{k}{N} = k \times \dfrac{1}{N}$.

Is $V \propto \dfrac{1}{N}$?

Table of values of V against $\dfrac{1}{N}$

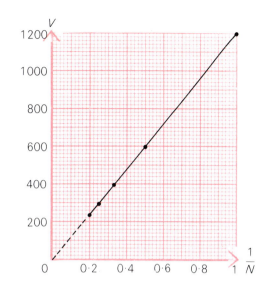

$\dfrac{1}{N}$	1	0·5	0·33	0·25	0·5
V	1200	600	400	300	240

The graph is a straight line through O, so
$V \propto \dfrac{1}{N}$, i.e. $V = \dfrac{k}{N}$ or $VN = k$.
From the table, $k = 1200$, so the relation is
$V = \dfrac{1200}{N}$.

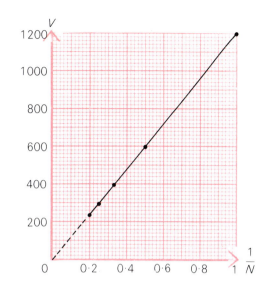

V is *inversely proportional* to N, or V *varies inversely* as N.

$$V \propto \dfrac{1}{N}, \ V = \dfrac{k}{N}, \text{ or } VN = k.$$

Exercise 5

1A Bluebell Rovers' sponsors decide to reduce the 'Spot the Ball' prize-money to £600.
   a Copy and complete these tables:

No. of prizes (N)	1	2	3	4	5
Value of each (£V)	600	300	200		

$\dfrac{1}{N}$	$\dfrac{1}{1} = 1$	$\dfrac{1}{2} = 0.5$	$\dfrac{1}{3} = 0.33$	$\dfrac{1}{4} = \ldots$	
V	600	300	200		$\ldots$

   b Draw a graph for each, like the ones above, but take 2 cm to £100 on the vertical scale.

   c Copy and complete: Since the graph of V against $\dfrac{1}{N}$ is a straight line through O,

     $V \propto \ldots\ldots$, or $V = \dfrac{k}{\ldots}$, or $VN = k$.

   d Find a formula connecting V and N.

PROPORTION IN PRACTICE

2A

What happens to the model train's acceleration when it has different loads to pull (see the graph)?

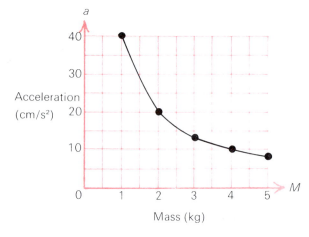

Mass (M kg)	1	2	3	4	5
Acceleration (a cm/s²)					

a Copy the table, and use the graph to complete it.

b Find the formula connecting a and M.

c Graph a against $\dfrac{1}{M}$. What do you find?

d Copy and complete: 'The acceleration is . . . .'

3A Show that for each journey the time taken is inversely proportional to the average speed, that is $T \propto \dfrac{1}{S}$, or $T = \dfrac{k}{S}$, or $ST = k$ (check every pair of values). Find k in each case.

a

S km/h	400	150	120	75	60
T h	$1\frac{1}{2}$	4	5	8	10

b

S m/s	0·5	0·8	1·0	1·25	2
T s	200	125	100	80	50

4A

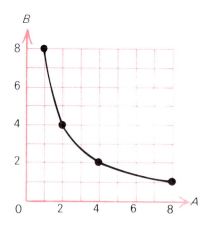

B varies inversely as A.

a Write this as an equation with a constant k in it.

b Choose a point on the graph, and use it to find k.

c What shape is the graph of B against $\dfrac{1}{A}$?

5A The current (I amps) in a circuit is inversely proportional to the resistance (R ohms). The current is 2 amps when the resistance is 250 ohms.

    a Find a formula for I.   **b** Calculate: (i) I when $R = 200$ (ii) R when $I = 4$.

6A The time (H hours) taken to deliver a batch of leaflets in the town centre varies inversely as the number (N) of people delivering them. For one job 60 people take 8 hours.

    a Find a formula for H.   **b** Calculate: (i) H when $N = 80$ (ii) N when $H = 12$.

7B The resistance (R ohms) to the current passing through a copper wire is inversely proportional to the square of the diameter (d mm) of the wire. For a wire of diameter 0·3 mm the resistance is 0·2 ohm. Find the formula connecting R and d.

8B Jake Maloney of Quickclear Motors, recorded the resale values of the Snorter Special Superbike:

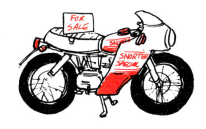

Age (A years)	1	2	3	4	5	6
Value (£V)	4000	2830	2310	2000	1790	1630

In fact, unknown to Jake, there is a good mathematical model for his table of values.

a Graph V against A. Looks hopeful . . . $V \propto \dfrac{1}{A}$? Check $V \times A$ in the table entries. Square one again?

b Try $V \propto \dfrac{1}{A^2}$, $V \propto \dfrac{1}{\sqrt{A}}$. Any use?

c What *is* the connection between V and A?

9B

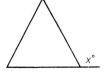 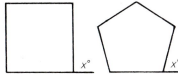

A slow puncture in the front tyre! The pressure is P after T minutes. Which of these models would you reject?

a $P \propto T$ **b** $P \propto T^2$ **c** $P \propto \dfrac{1}{T}$ **d** $P \propto \dfrac{1}{\sqrt{T}}$.

In fact, after 9 minutes, $P = 12$, and after a *further* 16 minutes $P = 7 \cdot 2$. Now choose the best model, and find the formula for P.

1 a $y \propto x$. Does it follow that $x \propto y$? Make a formula (with a variation constant) and find out.

b What about $y \propto \dfrac{1}{x}$? And $y \propto x^2$?

2

a Copy and complete the table for the sequence of regular polygons:

Number of sides (n)	3	4	5	6	...
Number of degrees in each exterior angle (x)					...

b Find a formula for x in terms of n.

c Write out your formula like this: The number of degrees in an exterior angle of a regular polygon is . . . . . . proportional to . . . .

JOINT VARIATION

How long will the cassette tape run?

(i) The greater its length, the longer it will run. The time (Ts) varies directly as the length (L feet). $T \propto L$.

(ii) The slower the speed, the longer it will record.

The time (Ts) varies inversely as the speed (S inches/s). $T \propto \dfrac{1}{S}$.

T varies directly as L and inversely as S, or T varies *jointly* as L and $\dfrac{1}{S}$.

$$T \propto \frac{L}{S}, \text{ or } T = \frac{kL}{S}.$$

Example

A tape 450 feet long runs for 24 minutes at $3\frac{3}{4}$ inches/second.

a Find the formula for T.

b Calculate the time of running of a tape 1000 feet long at $7\frac{1}{2}$ inches/second.

a $\quad T = \dfrac{kL}{S}$

$\quad\quad 24 = \dfrac{k \times 450}{3 \cdot 75}$

$\quad\quad k = \dfrac{24 \times 3 \cdot 75}{450} = 0 \cdot 2$

So $\quad T = \dfrac{0 \cdot 2L}{S}$.

b $T = \dfrac{0 \cdot 2L}{S}$

$\quad = \dfrac{0 \cdot 2 \times 1000}{7 \cdot 5}$

$\quad = 26\frac{2}{3}$

It runs for 26 minutes 40 seconds.

Exercise 6

1A A varies directly as r and as h. So $A \propto rh$, or $A = krh$. $A = 30$ when $r = 2$ and $h = 5$.
  a Find the formula for A in terms of r and h.
  b Calculate A when $r = 4$ and $h = 3$.

2A p varies directly as q and as r. $p = 12$ when $q = 2$ and $r = 3$.
  a Find the formula for p in terms of q and r.
  b Calculate p when $q = 1$ and $r = 5$.

3A V varies directly as x, y and z. $V = 80$ when $x = 2$, $y = 4$ and $z = 5$. Find the formula for V in terms of x, y and z.

4A I varies directly as m and as the square of r. $I = 100$ when $m = 5$ and $r = 2$.
  a Find the formula for I in terms of m and r.
  b Calculate I when $m = 8$ and $r = 3$.

5A Write a formula for each of these:
   a The power (E) in a circuit is directly proportional to the square of the current (C) and the length (L). (Remember the constant.)
   b The stopping distance (D) of a train is directly proportional to the square of its speed (S), and inversely proportional to the resistance (R) to its motion.
   c The distance (D) possible on a full tank of petrol varies directly as the capacity (C) of the tank, and inversely as the square root of the speed (S).

6A y varies directly as x and inversely as z. So $y \propto \dfrac{x}{z}$, or $y = \dfrac{kx}{z}$. $y = 3$ when $x = 2$ and $z = 4$.

   a Find the formula for y in terms of x and z.
   b Calculate y when $x = 5$ and $z = 10$.

7A F varies directly as m and inversely as n. $F = 50$ when $m = 25$ and $n = 8$. Find the formula for F in terms of m and n.

8A A varies directly as s and inversely as t^2. $A = 60$ when $s = 20$ and $t = 5$.
   a Find the formula for A in terms of s and t.
   b Calculate A when $s = 9$ and $t = 9$.

9B Write these formulae in words, using some of the following:
   directly proportional, inversely proportional, varies directly, varies inversely, varies jointly:

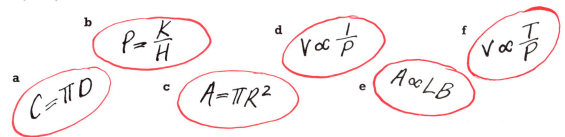

a $C = \pi D$
b $P = \dfrac{K}{H}$
c $A = \pi R^2$
d $V \propto \dfrac{I}{P}$
e $A \propto LB$
f $V \propto \dfrac{T}{P}$

=========== *Exercise 7* ===========

1A Safe and Secure Insurance Company provides pensions (£P) for its employees which are directly proportional to their length of service (Y years) and to their final salary (£S). Mrs Jackson had 20 years service and a final salary of £15 000. Her pension is £3750.
   a Calculate the formula for P.
   b Calculate the pension for 30 years service and a salary of £18 000.

2A The simple interest (£I) in a bank account varies directly as the principal (£P), the time (T years) and the rate of interest p.a. ($R\%$). The interest on £800 for 6 months (0·5 year) at 5% p.a. is £20.
   a Find the formula for I.
   b Calculate the interest on £1600 for 9 months at 7% p.a.

3A The volume (V cm³) of a certain mass of gas varies directly as the temperature ($T°$) and inversely as the pressure (P mm Hg)
   $V = 200$ when $T = 250$ and $P = 750$.
   a Find a formula for V in terms of T and P.
   b Calculate V when $T = 350$ and $P = 1000$.

PROPORTION IN PRACTICE

As the Daredevil Drum at the fairground speeds up you're held against the side. The force on you (F newtons) varies directly as the square of your speed (S m/s) and your mass (M kg), and inversely as the radius of the drum (R m).

a Given $F = 20$, $S = 10$, $R = 5$ when $M = 75$, find the formula for F.

b Calculate the force on Craig, whose mass is 50 kg, when the speed is 15 m/s.

5B The volume (V cm³) of a cone varies directly as the square of its radius (r cm) and its height (h cm). For radius 10 cm and height 30 cm the volume is 3142 cm³, to the nearest cm³.

a Find a formula for V, and calculate V when $r = 5$ and $h = 12$.

b What happens to the volume when:

(i) h is doubled   (ii) r is doubled   (iii) h and r are both doubled?

6B Miss Temple is an accountant with Allround Communications. She reports to the Board of Directors that, with a workforce of 150, and £250,000 spent on advertising, the year's profit is £500 000. She has found that the profit varies directly as the square root of the advertising spending and inversely as the number of employees.

a Use her model to make a profit formula.

b The Board decides to give 30 employees early retirement, and wants to aim for £1·5 million profit. How much would Miss Temple advise for advertising?

c If the number of employees and the profit both increase by 10%, calculate the percentage change in advertising costs.

7B Astrid knows that one way to 'lose' weight is to travel in space! In fact her weight is inversely proportional to the square of her distance from the centre of the Earth. A weight of W kg on Earth, radius R km, becomes w kg at a distance of d km from the centre.

a Write down formulae for:

(i) W in terms of R

(ii) w in terms of d.

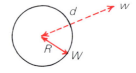

b Prove that $\dfrac{w}{W} = \dfrac{R^2}{d^2}$.

c Astrid weighs 80 kg on Earth. Calculate her weight 3200 km *above* the surface. (Radius of Earth = 6400 km.)

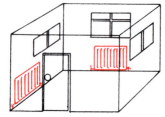

1 Imagine that you are one of H. H. Hothouse's central heating engineers. What factors would you consider before deciding how many radiators to recommend for a room?

Think about the various things that would determine the time taken to 'heat' the room. Construct a possible formula that could be used as a mathematical model for all of H. H. Hothouse's advisers. Explain your results to a teacher or parent or friend, to find out if they can be understood.

2 The electrical power in a circuit (*P* watts) is directly proportional to the current (*I* amperes) and the voltage (*V* volts). For 1200 watts the current is 5 amps and the voltage is 240. Find the formula for *P*. Investigate the current, and hence the size of fuse in the plug, for different household items such as a hair-drier (500 W), iron (1000 W), fire (3 kW), etc.

3

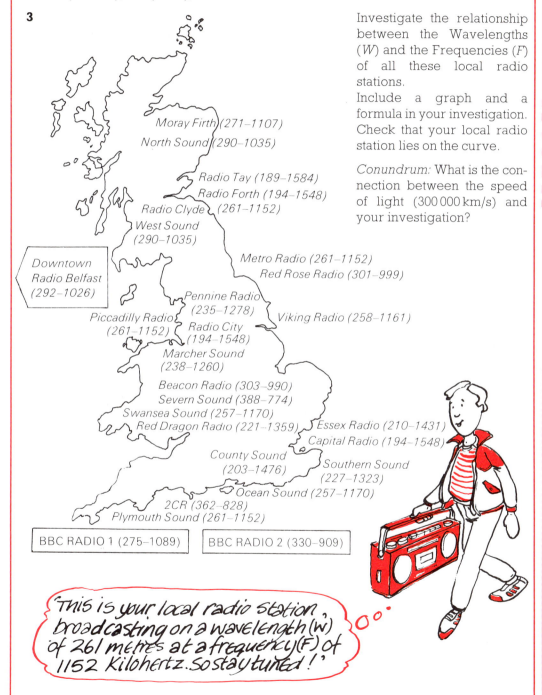

Moray Firth (271–1107)
North Sound (290–1035)
Radio Tay (189–1584)
Radio Forth (194–1548)
Radio Clyde (261–1152)
West Sound (290–1035)
Downtown Radio Belfast (292–1026)
Metro Radio (261–1152)
Red Rose Radio (301–999)
Pennine Radio (235–1278)
Piccadilly Radio (261–1152)
Viking Radio (258–1161)
Radio City (194–1548)
Marcher Sound (238–1260)
Beacon Radio (303–990)
Severn Sound (388–774)
Swansea Sound (257–1170)
Red Dragon Radio (221–1359)
Essex Radio (210–1431)
Capital Radio (194–1548)
County Sound (203–1476)
Southern Sound (227–1323)
Ocean Sound (257–1170)
2CR (362–828)
Plymouth Sound (261–1152)
BBC RADIO 1 (275–1089)   BBC RADIO 2 (330–909)

Investigate the relationship between the Wavelengths (*W*) and the Frequencies (*F*) of all these local radio stations.

Include a graph and a formula in your investigation. Check that your local radio station lies on the curve.

Conundrum: What is the connection between the speed of light (300 000 km/s) and your investigation?

'This is your local radio station, broadcasting on a wavelength (w) of 261 metres at a frequency (F) of 1152 Kilohertz. So stay tuned!'

1 p is proportional to the square root of q.
  a Find the formula connecting p and q.
  b Copy and complete the table.

q	64	100	
p	6		10

2 M varies inversely as N. $M = 12$ when $N = 6$. Find:
  a The formula for M **b** M when $N = 3$ **c** N when $M = 16$.

3 Sketch a graph for the model: **a** $y = kx$ **b** $y = \dfrac{k}{x}$ ($k > 0$ in both).

4 $F \propto \dfrac{kv^2}{r}$, where k is a constant. Write this formula in words.

5  The weekly bonus (£B) for each worker at Sparks' factory is directly proportional to the weekly profit (£P). For the week ending 22nd June, £16 bonus is paid for a £20 000 profit.
  a Find a formula for B.
  b Calculate: (i) the bonus for £21 500 profit   (ii) the profit for an £18 bonus.

6 A farmer makes square sheep–pens with 2 m lengths of fencing.
  a Copy and complete:

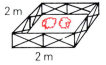

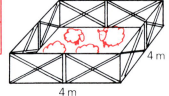

Number of 2 m fences per side (N)	1	2	3	4	5
Grazing area (A m²)	4	16			

  b Find a formula for A in terms of N.

7 The number of rectangular concrete slabs (N) required to pave a play area varies inversely as the area of each slab (A cm²). 192 slabs, measuring 90 cm by 60 cm, can cover the area.
  a Find a formula for N in terms of A.
  b How many slabs measuring 60 cm by 60 cm would be needed?
  c If A is doubled, what happens to N?

8 a Which of the following describes the relationship between x, h and V in the table?

$$V \propto xh \qquad V \propto \frac{h}{x^2} \qquad V \propto x^2h \qquad V \propto \frac{x^2}{h}.$$

x	1	2	3	4
h	30	15	10	7·5
V	10	20	30	40

  b Find the formula for V. What solid might this be a formula for if V stands for 'volume'?

BILLS

Running a house or car costs money. Could this be a scene from your home?

═══════════════ *Exercise 1* ═══════════════

1A Study these gas, telephone and electricity bills. Calculate the entries for the spaces A, B, C, etc, and so find out the total amount due in each one.

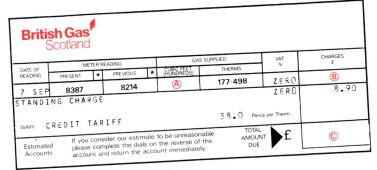

British Gas Scotland

DATE OF READING	METER READING				GAS SUPPLIED		VAT %	CHARGES £
	PRESENT	★	PREVIOUS	★	CUBIC FEET (HUNDREDS)	THERMS		
7 SEP	8387		8214		(A)	177·498	ZERO	(B)
STANDING CHARGE							ZERO	8·90
TARIFF CREDIT TARIFF					38·0 Pence per Therm			

Estimated Accounts — If you consider our estimate to be unreasonable please complete the dials on the reverse of the account and return the account immediately.

TOTAL AMOUNT DUE ► £ (C)

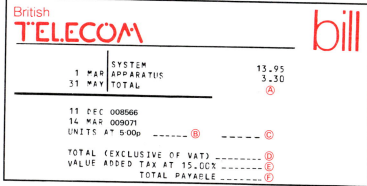

British **TELECOM** **bill**

1 MAR	SYSTEM	13·95
31 MAY	APPARATUS	3·30
	TOTAL	(A)

```
11 DEC 008566
14 MAR 009071
UNITS AT 5·00p  _____ (B)      _____ (C)

TOTAL (EXCLUSIVE OF VAT) _____ (D)
VALUE ADDED TAX AT 15·00% _____ (E)
           TOTAL PAYABLE _____ (F)
```

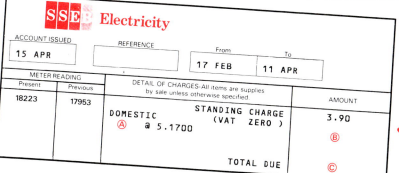

SSE Electricity

ACCOUNT ISSUED   **15 APR**   REFERENCE   From **17 FEB**   To **11 APR**

METER READING		DETAIL OF CHARGES·All items are supplies by sale unless otherwise specified.	
Present	Previous		AMOUNT
18223	17953	DOMESTIC STANDING CHARGE (VAT ZERO)	3·90
		(A) @ 5·1700	(B)
		TOTAL DUE	(C)

2A a Which bill is easiest to follow?
   b Can you suggest clearer layouts or instructions?

MORTGAGES

Are you buying a house? You'll almost certainly need a loan, called a *mortgage*.

Where can I get a mortgage?

From a Building society, a Bank or an Insurance company

usually between 80% and 100% of the value of the house. The rest you pay as your deposit.

How much can I get?

How do I repay the loan?

By a repayment mortgage, over an agreed number of years. Part of the loan is repaid monthly with interest. When the time is up, the loan has been repaid.

OR

By an endowment mortgage over an agreed number of years. Premiums for a life assurance policy and interest on the loan are paid monthly. When the time is up the policy matures and repays the loan.

How much interest do I have to pay?

This depends on the money markets. It is usually between 6% and 12%, including income tax relief.

=== Exercise 2 ===

1A

Betty and Jim are interested in buying a flat valued at £24 000. The Sandcastle Building Society offers them a loan for 80% of this value.

a How much will the Building Society lend?
b How much more will Betty and Jim have to find themselves to meet the price?

2A Mr Symon sees a house he would like to buy. It is valued at £65 000. The Union Jack Bank agrees to lend him 90% of the value. How much money does he have to find?

3A Jim Wright earns £215 a week. The Union Jack Bank offers him a loan of 3 × his annual salary for a house loan. How much is this?

4A
FOR SALE. Desirable flat. 2 public rooms and 2 bedrooms. Must be seen. Phone 100100.

Just what Iain and Astrid were looking for! They have it valued—£39 500. The Safety First Insurance Company offers them a loan of 95% of the value, or 2 × Iain's salary plus 1 × Astrid's salary—whichever is less. Iain's salary is £12 500 and Astrid's is £13 000. What size of loan could they have?

5A Copy and complete this calculation of the monthly repayments for an £18 000 mortgage over 15 years with the Safe and Secure Building Society.

Mortgage	Payment
£10 000 ⟷	£94·24
£18 000 ⟷	$£94·24 \times \dfrac{18\,000}{10\,000} = £\ldots$

Safe and Secure Building Society			
Monthly repayments on a £10 000 mortgage			
15 years	20 years	25 years	30 years
£94·24	£83·69	£76·81	£72·70

The monthly payment = £....

6A Calculate the monthly premiums with Safe and Secure for mortgages of:
   a £30 000 over 25 years   **b** £21 500 over 15 years.

7A Mr and Mrs Humphries hope to buy a house they see which is valued at £50 000. Safe and Secure offer them an 80% mortgage over 20 years. What would their monthly payment be?

8A *1% isn't much, is it?*

When the rate of interest charged by Safe and Secure rises by 1%, their monthly repayment charge on a £10 000 loan for 25 years increases to £77·58. How much more does a £10 000 borrower have to pay: **a** monthly   **b** annually?

9A Miss Li has a £24 000 mortgage for 20 years. She takes out a term assurance policy for the same amount at the same time. If she dies, this will repay her mortgage. The monthly premium is £4·16. How much does she pay altogether monthly for mortgage and assurance?

10B

It's a busy main road, but Fiona has set her heart on the flat at number 1128. The Safe and Secure Building Society values it at £31 800. They offer her a 20 year mortgage of 85% of the value, or $2\frac{1}{2}$ × her annual salary of £10 500, whichever is less.

   a How much will the Society lend her?   **b** Calculate Fiona's monthly repayment.
   c (i) How much will she pay over the whole period if the payments remain the same?
   (ii) How much of this is interest on the loan?

11B a Calculate the total payments to Safe and Secure for a loan of £10 000 over:
   (i) 15 years   (ii) 30 years.
   b What percentage of the total is interest in each case?

12B The rate of interest for a mortgage, before tax relief, is 15% per annum. Calculate the rate after 27% tax relief.

HOUSE AND CAR COSTS

HOUSE AND CONTENTS INSURANCE

Class discussion

If you buy a house with a mortgage you have to insure the house. When you rent or buy a house or flat it is wise to insure its contents also. Insurance policies cover loss or damage caused by fire, explosion, lightning, earthquake, storm, flood, impact by aircraft, vehicles, animals, aerials and trees, and, for contents, theft or damage by thieves. How could these two householders benefit by house and contents insurance?

Did you know? A house is broken into every four minutes in Britain.

═══════════════════ *Exercise 3* ═══════════════════

1A The Safety First Insurance Company charges an annual premium for house insurance of £1·50 per £1000 of the value of the house.

Copy and complete this calculation of the annual premium for a house valued at £36 000:

Value		*Premium*
£1000	⟷	£1·50
£36 000	⟷	£1·50 × $\dfrac{36\,000}{1000}$
		= £ . . . .

2A Calculate the annual premiums for house insurance charged by Safety First for:
  a a flat worth £20 000
  b a semi–detached villa valued at £45 000
  c a house worth £88 500.

3A Rebuilding a house can cost $1\frac{1}{2}$ times its market value, so the insurance company advises people to use this value for their house insurance calculation. How much would each premium in question **2A** become?

4A Safety First use this table to calculate annual premiums for every £1000 of house *contents* insured:

DISTRICT	A	B	C	D	E	F	G	H
Valuables £	11	12	15	18	23	34	39	46
Other items £	4	5	6	8	10	12	14	16

a Why are these rates much higher than for the buildings' insurance (£1·50 per £1000)?
b Which parts of Britain might correspond to: (i) District A   (ii) District H?
c List five common household *valuables*.

5A Steven has no valuables, but wants to insure the rest of his household belongings. He estimates that they are worth £8000. Calculate his annual premium (District G).

6A Sue and Harry have saved hard for years to buy the things in this room. They live in District D.
a Calculate their annual premium for insuring all these valuables.
b They reckon the other items in the house are worth insuring for £17 000. What premium have they to pay for these?

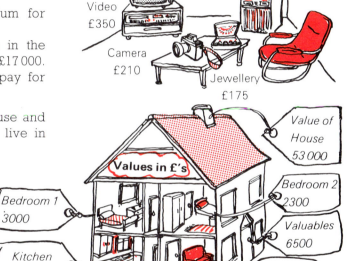

Music centre £1200
TV £420
Video £350
Camera £210
Jewellery £175

7B The King family insure their house and contents with Safety First. They live in District B. Calculate:
a their total annual premium
b the monthly premium if 10% is added for administrative costs.

Values in £'s

Value of House 53 000
Bedroom 1 3000
Bedroom 2 2300
Valuables 6500
Kitchen 5000
Lounge 5000

8B Many insurance companies keep the house and contents values up-to-date by *index–linking* them. This means that they increase the values *annually* in line with the rate of inflation. Taking an annual rate of inflation of 5%, calculate the values of this house and its contents after 1, 2 and 3 years: House—value £42 000   Contents—value £12 000.

---

1 *Either* **a** Choose a room in your own home. List the 'replacement' value of every item in it.
  Or **b** List the replacement value of all the *valuables* in your house.
    Are you surprised by the results?

2 Find out about the insurance of your house and its contents. Are your own belongings insured—bicycle, sports equipment, audio, and the rest?

RUNNING A CAR

Class discussion

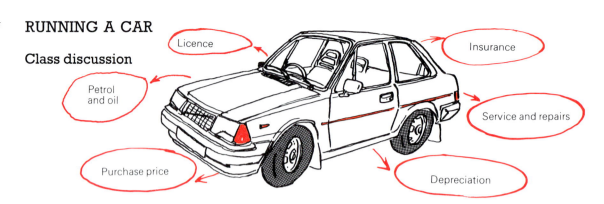

Licence

Insurance

Petrol and oil

Service and repairs

Purchase price

Depreciation

═══════════════ *Exercise 4* ═══════════════

1A Anne and Peter have a new car worth £6000. They want to find out how much it will cost them in a year. They estimate it will cover 10 000 miles at 35 miles to the gallon of petrol, and they assume that a gallon will cost 175p.

  a (i) How many gallons of petrol would they use?
      (ii) What would the cost of petrol be for the year?

  b Copy and complete:

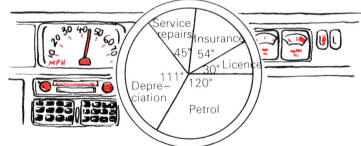

Licence	£ 100
Insurance	240
Service and repairs	80
Petrol	

      Total: £   .   .   .   .   *Cost per mile = . . . p*

Depreciation
= 20% of car's value = £   .   .   .   .   .   .   *Cost per mile = . . . p*

      Total: £   .   .   .   .   .   *Total cost per mile = . . . p*

2A Calculate the costs in the same way for a car worth £5000, covering 8000 miles in the year at 40 miles per gallon. The cost of insurance is £180, and service and repairs total £90. Depreciation is 15% of the car's value, and monthly Hire Purchase payments of £75 have to be made.

3A Two years later, Anne and Peter's car costs looked like this. The licence cost £110.

  a Make a list of the items and their cost.
  b What was the total cost for the year?

Service repairs 45°   Insurance 54°   Licence 30°
Depreciation 111°   Petrol 120°

Investigate the cost of travel by car, 'plane and train.

CAR INSURANCE

Class discussion

What do you know about these?

Third Party, Fire and theft

No-claims bonus

comprehensive car insurance

£50 EXCESS

under 25

The Banger Car Insurance Company charges these annual premiums for comprehensive car insurance:

Car group	District					
	A	B	C	D	E	F
1	225	240	265	285	310	340
2	270	285	310	340	355	400
3	315	340	365	400	425	470
4	380	405	440	480	500	565
5	420	465	505	545	580	660
6	515	540	590	650	695	770
7	620	675	730	790	840	930

Under 25 years of age, add 10%.
No–claims bonus: $33\frac{1}{3}$% discount if no claim in previous year
40% discount if no claim in previous 2 years
50% discount if no claim in previous 3 years
60% discount if no claim in previous 4 years or over.

Exercise 5

1A a Why do the premiums differ from place to place?
  b Why are cars arranged in different groups for insurance?
  c Why is it compulsory for cars to be insured (unlike houses, for example)?

HOUSE AND CAR COSTS

2A Calculate the annual premium for each of the following:

Name	Car group	District	Age	Years without a claim
D. Barr	1	D	30	0
M. Munn	6	B	37	1
K. Burns	3	E	28	3
R. Dick	2	C	21	2
A. Breen	4	A	18	1
I. Best	7	F	25	8

3A

Ella Robinson drives a Mini 850 (group 1). She is over 25, and lives in London (district F).

a Calculate her monthly car insurance premium if she has had one year without a claim.

b After the accident she loses her no–claims discount, and buys a group 3 car. Calculate the *increase* in her monthly premium.

4B Mr Pinkerton is an owner-only driver. He gets 20% off the insurance premium, and his no–claims discount is then calculated on the reduced premium. How much does he pay to insure his group 5 car, in district D, if he has not made any claims in the past three years?

5B 'That's not so bad', thought Ian Best. 'My 60% discount would be 40% after a claim.' What would be the increase in his premium, following an accident claim? (See Ian's details in the table in question **2A**.)

Banger cars Insurance co.

'Following an accident claim, a 60% bonus becomes 40% and a 50% bonus becomes 30%'

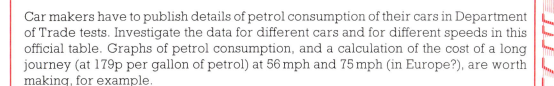
Car makers have to publish details of petrol consumption of their cars in Department of Trade tests. Investigate the data for different cars and for different speeds in this official table. Graphs of petrol consumption, and a calculation of the cost of a long journey (at 179p per gallon of petrol) at 56 mph and 75 mph (in Europe?), are worth making, for example.

DoT FUEL CONSUMPTION TESTS mpg (Litres/100km)			
	Simulated Urban Driving	Constant Speed	
		56mph (90km/h)	75mph (120km/h)
1300 Saloon (4-M)	32.5(8.7)	54.3(5.2)	42.0(6.7)
1300 Saloon (5-M)	31.7(8.9)	56.5(5.0)	43.6(6.5)
1300 Saloon (3-A)	30.0(9.4)	45.5(6.2)	36.7(7.7)
1600 Saloon (5-M)	33.2(8.5)	57.6(4.9)	42.8(6.6)
1600 Saloon (3-A)	32.5(8.7)	47.9(5.9)	35.8(7.9)
1600D Saloon (4-M)	44.8(6.3)	65.7(4.3)	46.3(6.1)
1600D Saloon (5-M)	42.8(6.6)	68.9(4.1)	48.7(5.8)
1800i Saloon (5-M)	28.5(9.9)	55.4(5.1)	43.5(6.5)
1800i Saloon (3-A)	27.2(10.4)	44.1(6.4)	35.8(7.9)

4-M = 4-speed transmission.   5-M = 5-speed transmission.
3-A = 3-speed automatic transmission.

APPRECIATION AND DEPRECIATION.

Houses tend to *appreciate*, or increase in value, year by year; cars usually *depreciate*, or decrease in value year by year. Why?

Example

Jim and Jenny bought a house for £36 000. In the following three years its value appreciated by 10%, 6% and 5%. Calculate its value each year.

First year   Value = £36 000
                          Appreciation = 10% of £36 000 = £3600
Second year   Value = £39 600
                          Appreciation = 6% of £39 600 = £2376
Third year   Value = £41 976
                          Appreciation = 5% of £41 976 = £2098
          Value after 3 years = £44 074.

Note   Another way: The value after 1 year is 110% of the original value = 1·1 × £36 000 = £39 600, and so on.

1A Mr and Mrs Jackson bought a flat for £32 000. In each of the following two years its value appreciated by 8%. How much was it worth at the end of two years?

2A

Jock's car *depreciated* rapidly. It cost him £4000.
a The first year it lost 25% of its value. What was it worth then?
b The second year it lost 20% of this value. What was it worth then?
c The third year it lost 15% of this value. What was it worth at the end of the three years?

3A Calculate the percentage appreciation of the value of this house:
a over the first year
b over the three year period.

| 1985 | 1986 | 1987 | 1988 |
| £45 000 | £47 250 | £49 140 | £53 070 |

4A

£3500   £2700   £2300   £2000

Calculate the percentage depreciation in this car's value:
a over the first year
b over the three year period.

5A The value of the machinery in a factory depreciates by 7% annually. The machinery cost £960 000 new. Calculate its value after two years.

6A Mr Richman valued his shares at the beginning of each year. Use the graph to calculate the percentage change in their value each year, compared to the previous year's value.

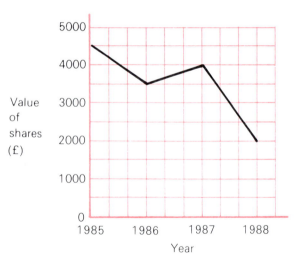

7B To find the *real* appreciation in value of Jim and Jenny's house in the *Example* on page 119, the annual rate of inflation has to be taken into account.
If the rate of inflation was 5% each year, the real appreciation in the first year was $(10-5)$%, that is 5% of £36 000.
Calculate the real appreciation each year in this way.

8B The value of Trisha's coin collection fell by 5%, to £430. Calculate its previous value. (Take this to be 100% of the value.)

9B The graph shows how the values of the peano in Moldrovia and the pashma in Kashmalia compared with the British pound in 1987.

   a When were the peano and pashma: (i) strongest (ii) weakest against the pound?

   b Calculate the percentage appreciation or depreciation in 1987 of:

     (i) the peano relative to the pound (ii) the pashma relative to the pound.

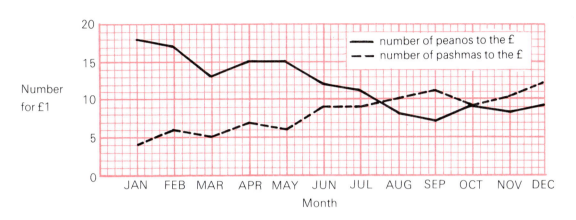

Every month the value of the Retail Price Index (RPI) is published. Investigate how this index is calculated. Show graphically how it has varied during the past year, or since it started at 100 in January 1974. Why is it a useful National Statistic?

CHECK-UP ON **HOUSE AND CAR COSTS**

1 **Mortgages**

The monthly repayment per £10 000 mortgage over 20 years with the Best of Brit Building Society is £85·76. The Wilsons hope to buy a house costing £47 500. Calculate the monthly repayment on a 95% loan.

2 **House and Contents Insurance**

The Rest Easy Insurance Company insures Jim Scott's valuables for £1·40 per cent, and his other house contents for £0·74 per cent. His house is insured for $1\frac{1}{2}$ times its value at £0·18 per cent. Calculate Jim's total annual insurance premium for:

valuables worth £3500, other contents worth £11 000 and house valued at £38 000.

Car Insurance

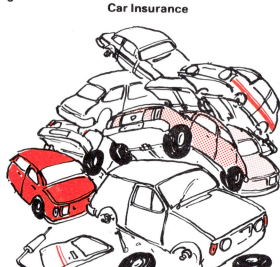

Imp Insurance charges these annual premiums for comprehensive car insurance in its area.

Car Group						
1	2	3	4	5	6	7
256	268	284	304	329	368	406

No claims bonus—$33\frac{1}{3}\%$, 40%, 50%, 60% discount for 1, 2, 3, 4 or more years without claims.

John drives a group 5 car, and has been accident-free for two years.

a Calculate his monthly premium.

b After an accident he loses his discount, and buys a group 1 car. What is the increase or decrease in his monthly premium?

4 **Appreciation and Depreciation**

a A house valued at £37 000 in 1986 was valued at £41 500 a year later. What was the percentage appreciation in its value?

b A company's computers depreciate by 12% annually; that is, at the end of each year their value is 12% less than at the beginning of the year. New, they cost £1 500 000. Calculate their value after 3 years.

Class discussion

1 Electrical power generated can be calculated using the formula $P = 2c^2$, where P is the power and c is the current.

If $P = 18$, $18 = 2c^2$,

$$\text{so } \boldsymbol{c^2 = 9}.$$

2 The area of the football pitch can be calculated using the formula $A = x(x+20)$.

If $A = 12\,000$, $x(x+20) = 12\,000$,

$$\text{so } \boldsymbol{x^2 + 20x - 12\,000 = 0}.$$

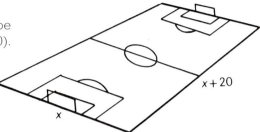

3 The distance travelled by the train can be calculated using the formula $D = t^2 - 5t$.

If $D = 750$, $750 = t^2 - 5t$,

$$\text{so } \boldsymbol{t^2 - 5t - 750 = 0}.$$

4 The line $y = 2x + 1$ cuts the curve $y = 3x^2$ at points A and B. The x-coordinates of A and B are given by $3x^2 = 2x + 1$,

$$\text{or } \boldsymbol{3x^2 - 2x - 1 = 0}.$$

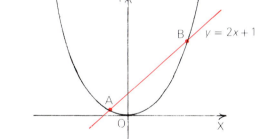

5 Purit, a company making air filters, has a total production cost £C for n filters, given by the formula
$C = -0{\cdot}0001n^2 + 10n + 10\,000$.

If $C = 100\,000$,
$100\,000 = -0{\cdot}0001n^2 + 10n + 10\,000$,

$$\text{so } \boldsymbol{0{\cdot}0001n^2 - 10n + 90\,000 = 0}.$$

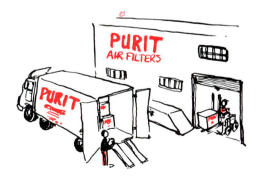

These are all *quadratic equations*, containing second degree terms like c^2, x^2, t^2 and n^2. Quadratic equations occur widely in mathematics and its applications.

HOW TO SOLVE QUADRATIC EQUATIONS   I. USING GRAPHS

QUADRATIC EQUATIONS

Guy sets off a rocket. Its height h metres after time t seconds is given by the formula $h = 18t - 3t^2$.

Using this table of values, a graph of h against t can be drawn.

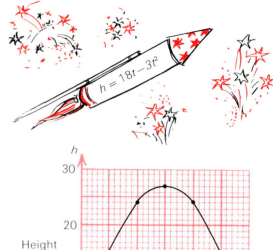

t	0	1	2	3	4	5	6
$18t$	0	18	36	54	72	90	108
$-3t^2$	0	-3	-12	-27	-48	-75	-108
h	0	15	24	27	24	15	0

The rocket takes off at $t = 0$, and lands at $t = 6$. At these times, $h = 0$.
So the solution of the equation $18t - 3t^2 = 0$ is $t = 0$ and $t = 6$. Check this by substitution.

The function f defined by $f(t) = 18t - 3t^2$ is a *quadratic function*. $18t - 3t^2 = 0$ is a *quadratic equation* with *solution*, or *roots*, 0 and 6.

=== *Exercise 1* ===

1A

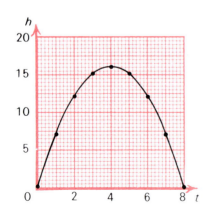

The graph is drawn of the flight formula $h = 8t - t^2$. Write down:
a the values of t when the rocket:
  (i) takes off   (ii) lands
b the solution of the quadratic equation $8t - t^2 = 0$.

2A a Copy and complete this table of values, until the rocket lands!

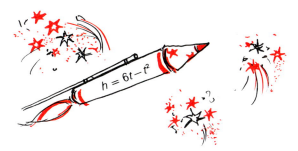

t	0	1	2	
$6t$	0	6		...
$-t^2$	0	-1		
h	0	5		

b Draw the graph of $h = 6t - t^2$ on 2 mm squared paper. Write down:
  (i) the values of t when the rocket takes off and lands
  (ii) the solution of the quadratic equation $6t - t^2 = 0$.

3A This rocket is already in the air when first seen.

t	-2	-1	
10	10	10	...
$3t$	-6	-3	
$-t^2$	-4	-1	
h	0	6	

a Copy and complete the table of values until the rocket lands.
b Draw the graph of $h = 10 + 3t - t^2$ on 2 mm squared paper.
c Write down:
  (i) the values of t when $h = 0$
  (ii) the solution of the quadratic equation $10 + 3t - t^2 = 0$.

4A The graph of $h = 8t - t^2$ again. To find when the rocket reaches a height of 12 metres, follow this.

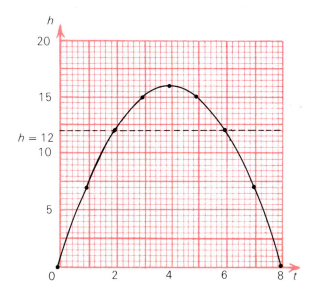

The line $h = 12$ is drawn (dotted). Read the values of t where this line cuts the curve. These are the solution of the equation

$$8t - t^2 = 12, \text{ i.e. } t^2 - 8t + 12 = 0.$$

5A a Place your ruler across the graph in question **4A** to solve these equations:
  (i) $8t - t^2 = 16$ (ii) $8t - t^2 = 15$ (iii) $8t - t^2 = 8$ (iv) $8t - t^2 = 0$.
  b Write each equation in the form $t^2 \ldots \ldots = 0$.

6A a Draw the graph of the parabola $y = x^2 - 8x + 7$, from $x = 0$ to 8.
  b Use it to solve these equations. (You'll need your ruler again.)
  (i) $x^2 - 8x + 7 = 0$ (ii) $x^2 - 8x + 7 = -5$ (iii) $x^2 - 8x + 7 = 7$.
  c Write the equations in **b**(ii) and (iii) in the form $x^2 \ldots \ldots = 0$.

7B a Draw the graph of the parabola $y = 21 + 4x - x^2$, from $x = -3$ to 7.
  b Use it to solve the equations:
  (i) $21 + 4x - x^2 = 0$ (ii) $21 + 4x - x^2 = 21$ (iii) $4x - x^2 = 4$.
  c Write each equation in the form $x^2 \ldots \ldots = 0$.

HOW TO SOLVE QUADRATIC EQUATIONS   II. USING FACTORS

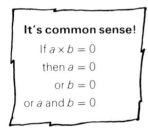

It's common sense!
If $a \times b = 0$
then $a = 0$
or $b = 0$
or a and $b = 0$

Example 1   Solve $8t - t^2 = 0$
$$8t - t^2 = 0$$
$$t(8 - t) = 0$$
$$\text{So } t = 0 \text{ or } 8 - t = 0$$
i.e. $t = 0$ or $t = 8$ (Compare with question **1A** of Exercise **1**.)

Example 2   Solve $(5 - t)(2 + t) = 0$
$$5 - t = 0 \text{ or } 2 + t = 0$$
So $t = 5$ or $t = -2$ (Compare with question **3A** of Exercise **1**.)

Exercise 2

Solve:
1A $x - 5 = 0$ **2A** $3 - x = 0$ **3A** $4x = 0$ **4A** $3x + 2 = 0$.

Find the two values of x in the solution of each equation:

5A $(x - 1)(x - 2) = 0$ **6A** $(x + 4)(x - 7) = 0$ **7A** $x(x - 8) = 0$

8A $3x(x + 5) = 0$ **9A** $(x + 3)(x - 3) = 0$ **10A** $x(2x - 1) = 0$

Solve:

11A $(x + 6)(2x - 3) = 0$ **12A** $6q(1 - 3q) = 0$ **13A** $(3r + 4)(5r - 8) = 0$

14A $(x - 1)^2 = 0$ **15A** $(2y + 1)^2 = 0$ **16A** $(4t + 1)(2t + 3) = 0$.

Find the coordinates of the points where these parabolas meet the x–axis:

17B $y = (x - 1)(2x - 9)$ **18B** $y = x(5 - x)$ **19B** $y = (3x - 2)(2x + 5)$.

1 A quadratic function f is defined by $f(x) = (x+2)(x-4)$.
  a Solve the equation $f(x) = 0$.
  b Calculate $f(0)$.
  c Write down the equation of the axis of symmetry of the graph of $f(x)=(x+2)(x-4)$.
  d *Sketch* the graph on plain paper, showing the coordinate axes and the axis of symmetry.
2 Repeat **1** for $f(x) = (x-2)^2$.

Reminders about factors

Exercise 3

Common factors

1A Factorise these. For example, $6x^2 - 18 = 6(x^2 - 3)$.
  a $3x^2 + 6$ **b** $5y^2 - 10$ **c** $6z^2 + 8$ **d** $4 - 2k^2$ **e** $12 - 12m$
  f $a^2 + 3a$ **g** $b^2 - 5b$ **h** $x^2 + x$ **i** $2d^2 + 4d$ **j** $8y - 12y^2$

Difference of squares

2A Factorise these. For example, $a^2 - b^2 = (a-b)(a+b)$.
  a $x^2 - y^2$ **b** $p^2 - q^2$ **c** $m^2 - 4$ **d** $x^2 - 16$ **e** $k^2 - 1$
  f $4x^2 - 1$ **g** $9y^2 - 4$ **h** $16 - z^2$ **i** $y^2 - 100$ **j** $16c^2 - 25$

Quadratic expressions

3A Factorise these. For example, $x^2 - x - 2 = (x-2)(x+1)$.
  a $x^2 + 3x + 2$ **b** $x^2 + 7x + 10$ **c** $y^2 + 2y + 1$ **d** $y^2 + 6y + 9$
  e $a^2 - 2a + 1$ **f** $b^2 - 8b + 16$ **g** $c^2 - c - 6$ **h** $d^2 + 3d - 10$
  i $2x^2 + 3x + 1$ **j** $3y^2 + 5y + 2$ **k** $2x^2 - 7x + 3$ **l** $8y^2 + 10y - 3$

A mixture (Look for a common factor first.)

4B Factorise:
  a $2x^2 - 8$ **b** $5y^2 - 5$ **c** $3a^2 - 3$ **d** $4c^2 - 36$
  e $4x^2 + 10x + 4$ **f** $8y^2 + 8y + 2$ **g** $2x^2 - 5x - 3$ **h** $6x^2 - 13x + 6$

Exercise 4

1A Copy and complete these solutions:

  a $x^2 + 5x = 0$ **b** $y^2 - 16 = 0$ **c** $2x^2 - 3x - 5 = 0$
  $x(\ldots) = 0$ $(y-4)(\ldots) = 0$ $(2x-5)(x+\ldots) = 0$
  $x = 0 \ or \ldots = 0$ $y - 4 = 0 \ or \ldots$ $2x - 5 = 0 \ or \ldots$
  $x = 0 \ or \ldots$ $y = 4 \ or \ldots$ $2x = 5 \ or \ldots$
  $x = 2\tfrac{1}{2} \ or \ldots$

  Solve:

2A $x^2 - 2x = 0$ **3A** $y^2 + y - 2 = 0$ **4A** $a^2 - 9 = 0$ **5A** $b^2 - 2b - 8 = 0$

6A $k^2 + 6k = 0$ **7A** $t^2 - 8t - 20 = 0$ **8A** $u^2 + 9u = 0$ **9A** $6x^2 - 6 = 0$

10A $2x^2 + 5x - 3 = 0$ **11A** $3y^2 - 10y + 3 = 0$ **12A** $9x^2 + 6x + 1 = 0$ **13A** $6y^2 - 5y + 1 = 0$

14A $2x^2 - 2x = 0$ **15A** $2x^2 - 18 = 0$ **16A** $2x^2 - 2x - 4 = 0$ **17A** $2x^2 = 0$

18B Find x where these parabolas meet the x–axis:

    a $y = 6x^2 - x - 2$ **b** $y = 4x - 6x^2$ **c** $y = 12x^2 - 17x + 6$ **d** $y = 2x^2 + 20x + 50$.

Explain the answer to **d**. Sketch one of the parabolas on plain paper. (Note that it cuts the y–axis at $x = 0$.)

QUADRATIC EQUATIONS

THE STANDARD FORM OF A QUADRATIC EQUATION

Every quadratic equation contains an x^2 term as the highest power of x.

> The *standard form* of a quadratic equation is $ax^2 + bx + c = 0$, $a \neq 0$.

To solve a quadratic equation, first arrange it in standard form.

Example 1 Solve $x(2x - 3) = -1$

$$2x^2 - 3x + 1 = 0$$
$$(2x - 1)(x - 1) = 0$$
$$x = \tfrac{1}{2} \text{ or } 1$$

Example 2 Solve $\dfrac{x}{4} - \dfrac{2x - 1}{x + 1} = 1$

Multiply each side by $4(x + 1)$

$$x(x + 1) - 4(2x - 1) = 4(x + 1)$$
$$x^2 + x - 8x + 4 = 4x + 4$$
$$x^2 - 11x = 0$$
$$x(x - 11) = 0$$
$$x = 0 \text{ or } 11$$

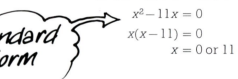

Exercise 5

Rearrange each equation in standard form, $ax^2 + bx + c = 0$, then solve it:

1A $x^2 + 5x = -4$ **2A** $x^2 - 2x = 8$ **3A** $x^2 - 7x = -6$ **4A** $x^2 + x = 12$

5A $x^2 = x$ **6A** $x^2 = 4$ **7A** $x^2 = 12x - 36$ **8A** $x^2 = -8x - 15$

9A $3x^2 + 3 = 10x$ **10A** $x^2 + 4 = 4x$ **11A** $1 + 6x^2 = 5x$ **12A** $4x^2 + 1 = 5x$

13A $x(x + 4) = 32$ **14A** $x(x - 5) = 24$ **15A** $4x(x + 1) = 15$ **16A** $12x - 4 = 9x^2$

17B $x + 2)(x + 3) = 6$ **18B** $(x + 1)^2 = 1$ **19B** $(2x - 3)^2 = 4$ **20B** $(x - 3)(2x + 3) = 5$

21B $\tfrac{1}{2}x(x + 1) = 10$ **22B** $x + \dfrac{2}{x} = 3$ **23B** $\tfrac{1}{2}x^2 - \tfrac{1}{3}x = \tfrac{1}{6}$ **24B** $2x = \dfrac{12}{x} + 5$

25B $\dfrac{6}{x} - \dfrac{6}{x + 1} = 1$ **26B** $\dfrac{1}{x - 2} - \dfrac{1}{x + 3} = \dfrac{1}{10}$ **27B** $\dfrac{3}{x + 1} + \dfrac{1}{x - 1} = 2$

Donald tried his own system for $(x-3)(x-5) = 8$.
He said $x-3 = 8$ or $x-5 = 8$.
    So $x = 11$ or 13.
What is wrong with his system?
Check that neither 11 nor 13 'satisfies' the equation.
Solve it correctly, and check your solution.

Graphs again

Example Find the coordinates of the points
where the line $y = x+1$ cuts the parabola
$y = x^2-5x+6$.
At the points of intersection, A and B,
$y = x^2-5x+6$ *and* $y = x+1$

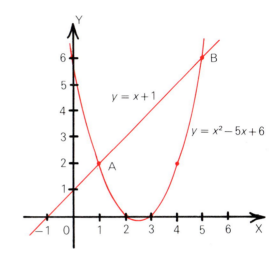

So $x^2-5x+6 = x+1$

$x^2-6x+5 = 0$

$(x-1)(x-5) = 0$

$x = 1$ or 5

$y = x+1 = 2$ or 6

A is the point $(1, 2)$ and B is $(5, 6)$.

=========== *Exercise 6* ===========

1A Find the coordinates of the points where each line cuts the parabola:
   a by reading them from the graph
   b by forming a quadratic equation, and solving it.

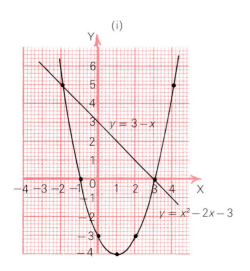

(i)

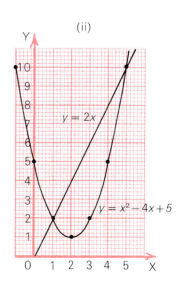

(ii)

2A A machine robot is programmed to cut along the line $y = x$ from O to A, then along the parabola to B, then back to O. Find the coordinates of the points it has to be given.

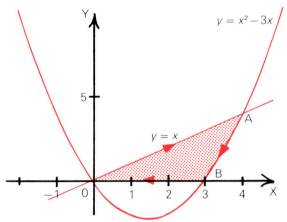

3B With the help of quadratic equations, find the coordinates of the points where these meet.

a The line $y = x - 4$ and the parabola $y = x^2 - 2x - 8$.

b The parabolas $y = x^2$ and $y = 8 - x^2$.

c The line $y = x$ and the cubic curve $y = x^3$.

QUADRATIC EQUATIONS AS MATHEMATICAL MODELS

Example Jim arrives with a $28\,m^2$ roll of Neverwear carpet for a *square* room. The carpet and the room have the same width. A 3 metre length of carpet is left over.

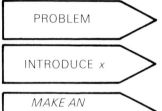

PROBLEM

Find the dimensions of the room.

INTRODUCE x

Let the floor be x m wide.

MAKE AN EQUATION

$$x(x + 3) = 28$$
$$x^2 + 3x - 28 = 0$$

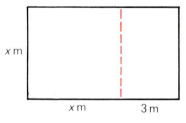

x m

x m   3 m

SOLVE THE EQUATION

$$(x - 4)(x + 7) = 0$$
$$x = 4 \text{ or } -7.$$

SOLVE THE PROBLEM

The length cannot be -7 metres. So the room is 4 m by 4 m.

═══ *Exercise 7* ═══

1A The length of this roll of carpet is 3 metres more than its breadth. Its area is $18\,m^2$. Find the length and breadth of the carpet. (Follow the signposts above.)

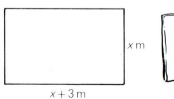

x m

Area = $18\,m^2$

$x + 3$ m

2A Another roll has an area of 60 m², and its length is 4 metres more than its breadth. Use a quadratic equation to find its length and breadth.

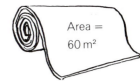

Area = 60 m²

3A

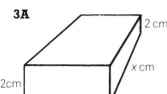

2 cm

x cm

2cm

$x + 1$ cm

The volume of the metal box is 40 cm³ Calculate the length and breadth of the box by making a quadratic equation and solving it.

4A When a clay pigeon is fired upwards, its height h metres after t seconds is given by the formula $h = 18t - 5t^2$.
  a How long does it take to reach a height of 9 m?
  b Why are there two answers?

5A

Up goes the flare to mark the start of a balloon flight. After t seconds the height of the flare is $50t - t^2$ metres, and the height of the balloon is $11t$ metres.
  a After how many seconds are they at the same height?
  b How can you tell that they start at the same height?

6A The height of the triangular gable end at 68 Orange Street is 5 metres more than the width of its base. The area of the triangle is 18 m². Calculate the height and width of the triangle.

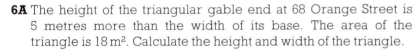

7A

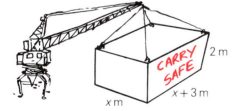

2 m

CARRY SAFE

$x + 3$ m

x m

Calculate the length and breadth of this goods container, given that its volume is 80 m³.

8A Apply Pythagoras' Theorem to this triangle to find the quadratic equation $x^2 + 2x - 48 = 0$. Use the equation to calculate the lengths of the two sides.

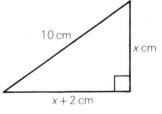

10 cm

x cm

$x + 2$ cm

9A

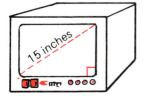

15 inches

The height of the screen on the television set is 3 inches less than the breadth. Taking x inches for the breadth, show that $x^2 - 3x - 108 = 0$. Hence find the dimensions of the screen.

10B Form a quadratic equation, solve it and then calculate the volume of the wedge.

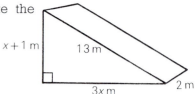

$x+1$ m   13 m   2 m $3x$ m

11B The sum, S, of n terms of $1+2+3+\ldots+n$ is given by the formula $S=\frac{1}{2}n(n+1)$.
   a Check the formula for $n=1, 2$ and 3.
   b Find n when $S=55$:   (i) by adding terms   (ii) using a quadratic equation.
   c Repeat **b** for $S=210$.

12B The closed container on the lorry is 3 m long, x m wide and x m high. Its *total surface area* is 32 m². Find x.

13B The number of units of length in the side of a square is equal to the number of units in its area. How many units are there in its length?

14B An aluminium strip 16 cm wide is bent to form a gutter. The area of the rectangular end is 32 cm².
   a Express the width of the end in terms of x.
   b Form a quadratic equation, and calculate the height and width of the gutter.

x cm

x cm

15B

Two cyclists cover 60 km at average speeds which differ by 5 km/h. One takes an hour longer than the other. Copy and complete the following to find their speeds.
Let the speeds be x km/h and $x+5$ km/h.

Their times are $\dfrac{60}{x}$ and $\dfrac{60}{\ldots}$ hours.

So $\dfrac{60}{x}-\dfrac{60}{\ldots}=\ldots.$

Multiply each side by $x(\ldots)$.

16B Repeat question **15B** for two motorists who travel 60 km at average speeds which differ by 10 km/h, one taking an hour longer than the other.

17B The cost (in £s) of producing x new home computers is $-5x^2+730x+2000$. They sell for £250 each. How many must be sold to 'break even'?

18B A cardboard pyramid on a rectangular base has volume 75 cm³ and height 5 cm. One side of the base is 4 cm longer than the other. Find the dimensions of the base. ($V=\frac{1}{3}$ area of base × height.)

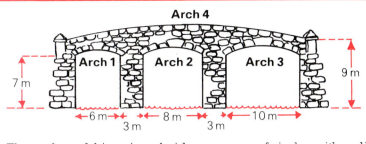

Arch 4

7 m   9 m

←6 m→ ←8 m→ ←10 m→
3 m   3 m

The arches of this unique bridge are arcs of circles with radii:
5 m for arches 1 and 2; 13 m for arch 3; 39 m for arch 4.

PROBLEM

Find the maximum clearance for arches 1, 2 and 3, and the maximum height of arch 4, above the river.

HINT

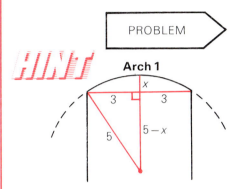

Arch 1

Using Pythagoras' Theorem,

$$(5-x)^2 + 3^2 = 5^2.$$

Go on to show that the maximum clearance above the river for Arch 1 is 8 m.
Then try the other parts of the problem.

---

a The diagram shows the graphs of the quadratic functions f, g and h defined by:

$f(x) = x^2 - 6x + 10$ (A)
$g(x) = x^2 - 6x + 9$ (B)
$h(x) = x^2 - 6x + 8$ (C)

How many solutions have the equations:

  (i) $f(x) = 0$
  (ii) $g(x) = 0$
  (iii) $h(x) = 0$?

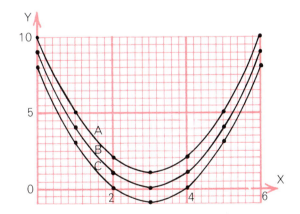

b Where does the graph of f cross the x–axis? Look at the equation $x^2 - 6x + 10 = 0$. $a = 1$, $b = -6$, $c = 10$. (Remember the standard form $ax^2 + bx + c = 0$.) Calculate the value of $b^2 - 4ac$. Repeat for curves B and C.

c Investigate the connection between the value of $b^2 - 4ac$ and the number of solutions of $f(x) = 0$, $g(x) = 0$ and $h(x) = 0$.

d Use the results of your investigation in **c** to classify the equations below, according to whether they have two different roots, two equal roots or no real roots.

  (i) $2x^2 - x - 3 = 0$ (ii) $4x^2 - 20x + 25 = 0$ (iii) $3x^2 - 2x - 4 = 0$ (iv) $x^2 - 4x + 5 = 0$.

HOW TO SOLVE QUADRATIC EQUATIONS
III. USING THE QUADRATIC FORMULA

The quadratic formula for solving the equation $ax^2 + bx + c = 0$ is

$$x = \frac{-b \pm \sqrt{(b^2 - 4ac)}}{2a}$$ $+$ for one root, $-$ for the other.

Example   Solve the equation $2x^2 - 5x - 1 = 0$

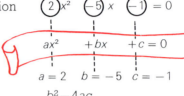

$$ax^2 \qquad + bx \qquad + c = 0$$

$$a = 2 \quad b = -5 \quad c = -1$$

$x = \dfrac{-b \pm \sqrt{(b^2 - 4ac)}}{2a}$, where

$= \dfrac{-(-5) \pm \sqrt{33}}{2(2)}$

$= \dfrac{5 + 5 \cdot 745}{4}$ or $\dfrac{5 - 5 \cdot 745}{4}$

$= 2 \cdot 69$ or $-0 \cdot 186$, to 3 significant figures.

$b^2 - 4ac$
$= (-5)^2 - 4(2)(-1)$
$= 25 + 8$
$= 33$

THE Equation Solver's collection of formulae.
Every formula guaranteed correct
All proofs given for each formula

=== *Exercise 8* ===

Give the solutions correct to 3 significant figures, where necessary.

1A Solve the equation $x^2 - 6x + 8 = 0$:   **a** using factors   **b** using the formula.
For **b**, compare $ax^2 + bx + c = 0$. $a = 1, b = -6, c = 8$.
$b^2 - 4ac = (-6)^2 - 4(1)(8) = \ldots$.

2A Solve $2x^2 - 8x - 3 = 0$, using the formula. Write down the values of a, b, c first. Then calculate $b^2 - 4ac$.

3A Use the quadratic formula to solve:
   a $x^2 + 2x - 1 = 0$ **b** $x^2 - 5x + 2 = 0$ **c** $3x^2 - 4x - 5 = 0$.

4A In each case solve the equation, and then explain the connection between its solution and the graph.

a
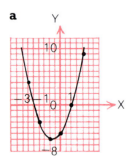

b
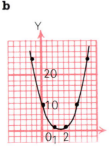

c
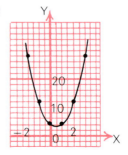

a Graph $y = 2x^2 + 3x - 5$.
Solve $2x^2 + 3x - 5 = 0$.

b Graph $y = 4x^2 - 12x + 9$.
Solve $4x^2 - 12x + 9 = 0$.

c Graph $y = 4x^2 - 4x + 5$.
Solve $4x^2 - 4x + 5 = 0$.

5A Take one card from each box, add them together and make a quadratic equation. For example, $2x^2 + 7x - 3 = 0$. Then solve the equation.
How many different equations are there?

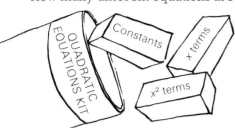 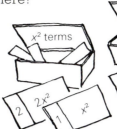  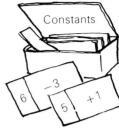

6B a Copy and complete the table of values for
  $y = 2x^2 + 3x - 7$ for $x = -4$ to 2.
  b Draw the graph of the equation on 2 mm squared paper.
  c Solve the equation $2x^2 + 3x - 7 = 0$:
   (i) using the graph (ii) using the formula.
  d Substitute the roots you get in the equation, and compare their accuracy.

x	-4	-3	...
$2x^2$	32		
$3x$	-12		
-7	-7	-7	
y	13		

IMPROVING APPROXIMATE SOLUTIONS

Give a computer an approximate root of an equation, and in a flash it will give you as accurate a value as you wish. Here is one of the methods it uses.

From the graph, approximate roots of the equation $x^2 + x - 4 = 0$ at A and B lie between -2 and -3, and between 1 and 2.
Find a better approximation for the root at B.

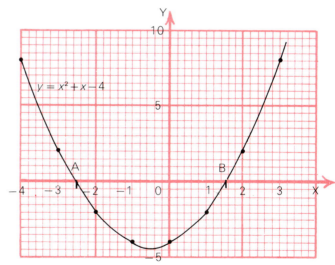

$y = x^2 + x - 4$

Step-by-step

If f is the function defined by $f(x) = x^2 + x - 4$, then
  $f(1) = 1^2 + 1 - 4 = -2$ (curve below x–axis)
  $f(2) = 2^2 + 2 - 4 = 2$ (curve above x–axis)
We want to find a, where $f(a) = 0$.
  $f(1 \cdot 5) = 1 \cdot 5^2 + 1 \cdot 5 - 4 \doteqdot -0 \cdot 25$ (below x–axis)
  $f(1 \cdot 6) = 1 \cdot 6^2 + 1 \cdot 6 - 4 \doteqdot 0 \cdot 16$ (above x–axis)

  $f(1 \cdot 55) = 1 \cdot 55^2 + 1 \cdot 55 - 4 \doteqdot -0 \cdot 05$
  $f(1 \cdot 56) = 1 \cdot 56^2 + 1 \cdot 56 - 4 \doteqdot -0 \cdot 01$
  $f(1 \cdot 57) = 1 \cdot 57^2 + 1 \cdot 57 - 4 \doteqdot 0 \cdot 03$.
The root is very close to $1 \cdot 56$ (actually $1 \cdot 5616$, correct to 4 decimal places).
From the symmetry of the graph, the other root is close to $-2 \cdot 56$.

QUADRATIC EQUATIONS

1B Show that the root of the equation $3x - 5 = 0$ lies between 1 and 2, and find a better approximation.
(Let $f(x) = 3x - 5$, and calculate $f(1)$ and $f(2)$.
Then try $f(1\cdot 6)\ldots$, etc.)

2B Prove that one root of the equation $x^3 - 3x + 1 = 0$ lies between 1 and 2, and find it, correct to 1 decimal place.
(You'll find your calculator useful, including the $\boxed{y^x}$ key.)

3B Repeat question **2B** for the root of the equation between 0 and 1.

4B a Use the graphs to estimate approximate solutions of the equations.
 b Use the step-by-step method to find the solutions correct to 1 decimal place.

(i)

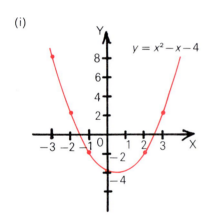

(ii)

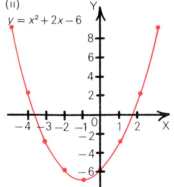

(iii)

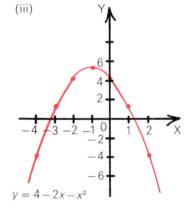

Solve $x^2 - x - 4 = 0$.   Solve $x^2 + 2x - 6 = 0$.   Solve $4 - 2x - x^2 = 0$.

 c Check your solutions by means of the quadratic formula.

5B One root of the equation $x^3 + x - 3 = 0$ lies between 1 and $1\cdot 5$. Find it, correct to 1 decimal place.

6B a Write down the equation which gives the points of intersection of the line $y = 2x$ and the parabola $y = x^2 - 6x + 11$, and simplify it.
 b One root of the equation lies between 6 and 7. Find it, correct to 2 decimal places.

1 Use the graph of $h = 6t - t^2$ to solve these equations
  a $6t - t^2 = 0$ **b** $6t - t^2 = 5$
  c $6t - t^2 = 9$ **d** $6t - t^2 = 8$
  e $6t - t^2 = 7$, correct to 1 decimal place.

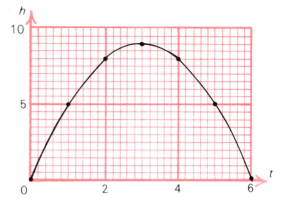

2 Solve:
  a $(x-1)(x+5) = 0$ **b** $y(y+1) = 0$ **c** $x^2 - 3x = 0$ **d** $y^2 - 25 = 0$
  e $(x+9)(2x+1) = 0$ **f** $x^2 - x - 6 = 0$ **g** $4y^2 - 4y + 1 = 0$ **h** $2y^2 - 7y - 4 = 0$.

3 Arrange in standard form, and solve:
  a $x^2 - 2x = 3$ **b** $y(y-4) = 21$ **c** $(3x-1)^2 = 4$ **d** $\frac{1}{2}x^2 = 9 - \frac{3}{2}x$.

4 Find the coordinates of the points of intersection of the straight lines and the parabolas:

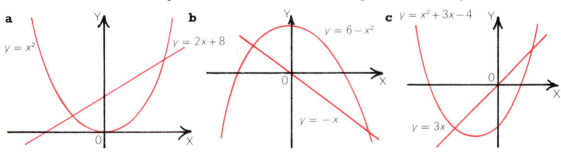

5 Make a quadratic equation for each, and find x:

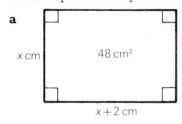 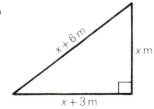 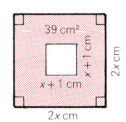

6 Take one card from each box. Add them together, make a quadratic equation and solve it (to 3 significant figures). Repeat for other sets of cards.

 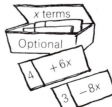 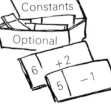

7 The curve $y = 2x^2 + 3x - 7$ cuts the x-axis between $x = 1$ and 2, and between $x = -2$ and -3. Use the step-by-step method to find the roots of $2x^2 + 3x - 7 = 0$ correct to 2 decimal places.

Investigation—live and learn

Ready to walk . . .

Ready to go solo . . .

Ready for the next level . . .

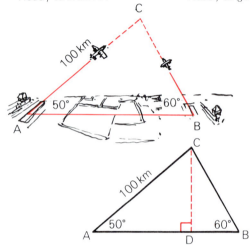

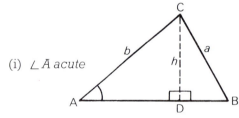

A fighter and a tanker aircraft fly from airfields at A and B, to rendezvous at C. How far has the tanker from B to fly?

a Is there enough information for a scale drawing? If so, draw one, and measure BC.

b Can you find BC by trigonometry? If not, why not?

c Try this. Sketch △ABC, and draw CD at right angles to AB. Calculate CD, and then BC. So you *could* use trigonometry, involving only sines of angles. Now read on.

THE SINE RULE

In any $\triangle ABC$, $\dfrac{a}{\sin A} = \dfrac{b}{\sin B} = \dfrac{c}{\sin C}$.

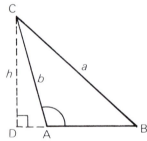

(i) ∠A *acute*

(ii) ∠A *obtuse*

Draw altitude CD to make two right-angled triangles, ADC and BDC. Let CD = h units.

In $\triangle ACD$,

$\sin A = \dfrac{h}{b}$ $\left(\text{In (ii), } \sin A = \sin(180° - A) = \dfrac{h}{b} \right)$

So $h = b \sin A$

In $\triangle BCD$,

$\sin B = \dfrac{h}{a}$

So $h = a \sin B$

Hence $b \sin A = a \sin B$

$\dfrac{b \sin A}{\sin A \sin B} = \dfrac{a \sin B}{\sin A \sin B}$ (dividing each side by $\sin A \sin B$).

So $\dfrac{a}{\sin A} = \dfrac{b}{\sin B}$ In the same way, $\dfrac{a}{\sin A} = \dfrac{c}{\sin C}$.

In any $\triangle$ABC, $\dfrac{a}{\sin A} = \dfrac{b}{\sin B} = \dfrac{c}{\sin C}$, the Sine Rule.

Example   Calculate c in $\triangle$ABC.

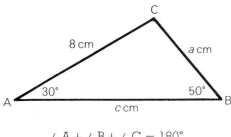

$\dfrac{a}{\sin A \checkmark} = \dfrac{b \checkmark}{\sin B \checkmark} = \dfrac{c}{\sin C \checkmark}$ (Tick the parts you're given.)

$\dfrac{8}{\sin 50°} = \dfrac{c}{\sin 100°}$

$c \sin 50° = 8 \sin 100°$

$c = \dfrac{8 \sin 100°}{\sin 50°} \left(= \dfrac{8 \sin 80°}{\sin 50°} \right)$

$= 10 \cdot 3$, to 1 decimal place.

$\angle A + \angle B + \angle C = 180°$
$\angle C = 180° - 30° - 50°$
$\quad = 100°.$

Note   **In this chapter give answers correct to 1 decimal place, unless there are other instructions.**

=================== Exercise 1 ===================

1A

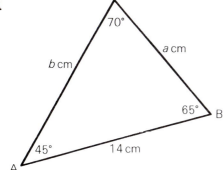

To calculate a, copy and complete:

$\dfrac{a}{\sin A} = \dfrac{b}{\sin B} = \dfrac{c}{\sin C}$ (Tick the side and angles you know.)

$\dfrac{a}{\ldots} = \dfrac{14}{\ldots}$

$a \sin \ldots = 14 \sin \ldots$

$a = \dfrac{14 \sin \ldots}{\ldots} = \ldots .$

2A Calculate a in each triangle:

a

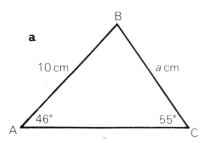

b

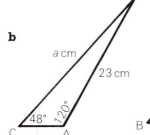

c

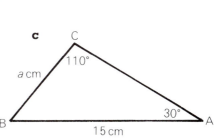

3A Sketch △ABC in which ∠A = 112°, ∠B = 51° and b = 18.
Calculate: **a** a **b** ∠C **c** c.

4A

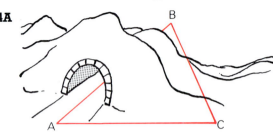

5A Meena sets sail at a point 25 m from the lighthouse. Her course is at an angle of 33° to the shoreline. Calculate, to the nearest metre: **a** PR **b** QR.

6A

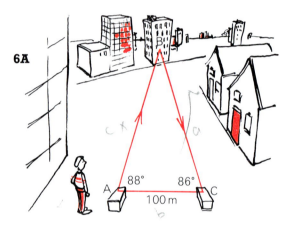

7B Observations of a hilltop T are made from points A and B, 100 m apart. The angles of elevation are 20° and 30°. Calculate:
a TB **b** the height of the hill.

8B

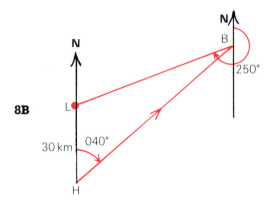

A tunnel is to be made along the line AB. To plan it, point C is chosen where A and B can both be seen. These measurements are then made:
∠A = 46°, ∠C = 68°, AC = 400 m.
Calculate, to the nearest metre:
a BC **b** AB.

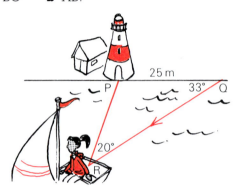

A laser beam is sent out from A. It bounces off the building at B and is picked up at C. Calculate, to the nearest centimetre:
a BA **b** BC.

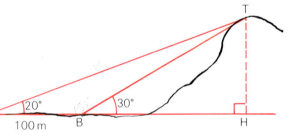

A ship leaves H and sails for 2 hours on a course bearing 040°, to B. From B, the lighthouse L bears 250°. Calculate:
a ∠HLB
b BH
c the average speed of the ship in km/h.

Calculating an angle

Example   Calculate $\angle$A and $\angle$B

$$\frac{a^{\checkmark}}{\sin A} = \frac{c^{\checkmark}}{\sin C^{\checkmark}}$$

$$\frac{3}{\sin A} = \frac{9}{\sin 120°}$$

$$9 \sin A = 3 \sin 120°$$

$$\sin A = \frac{3 \sin 120°}{9}$$

$$\angle A = 16 \cdot 8°$$

$$\angle B = 180° - 120° - 16 \cdot 8° = 43 \cdot 2°.$$

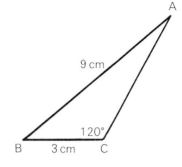

Note $\sin A = \sin (180° - A)$, so $\angle$A *could* be $180° - 16 \cdot 8° = 163 \cdot 2°$, by calculation.
From the given data in the triangle, $\angle$A must be acute, i.e. $16 \cdot 8°$. (Why?)

Exercise 2

1A Calculate the remaining two angles in each triangle:

a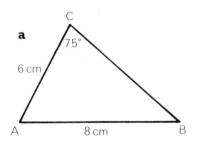

b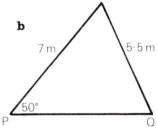

c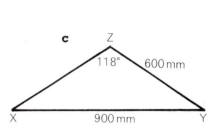

2A Sketch these triangles, and in each case calculate $\angle$A:
   a $\angle$C $= 68°$, $a = 28$ and $c = 35$ **b** $\angle$B $= 140°$, $b = 106$ and $c = 46$.

3A Habib measures his new bicycle.
   AB $= 16$ inches, AD $= 18$ inches, DC $= 25$ inches, $\angle$ABD $= 62°$ and $\angle$DBC $= 66°$.
   a Make a sketch of the bicycle frame.
   b Calculate: (i) $\angle$ADB (ii) DB (iii) $\angle$DCB.

4A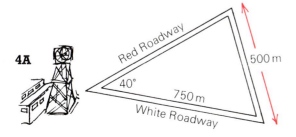

Underground in the Black Pit, the Red and White Roadways run at 40° to each other. Ted, a mining engineer, is planning a new 500 metre roadway link. Calculate the angle between the new link and the White Roadway.

TRIANGLES AND TRIGONOMETRY

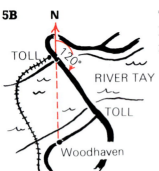

The Tay Road Bridge has tolls at either end, 2·5 km apart. The bridge runs on a bearing of 120°. Woodhaven is due South of the northern toll, and is 3 km from the southern toll.

a Draw the triangle formed by Woodhaven and the two tolls.

b Calculate the bearing of the southern toll from Woodhaven.

BRAINSFORMER

1

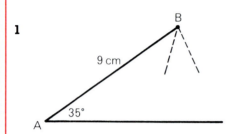

Given $\angle A = 35°$, $AB = 9$ cm and $BC = 6$ cm, make an accurate construction for two different positions of C on the base-line. Measure, then calculate, $\angle ACB$ in each case.

2 The heavy swinging weight is being used to demolish the building. Calculate:

a the angle of swing C_1BC_2,

b the length of the *arc* of swing from C_1 to C_2.

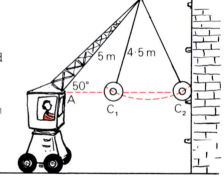

To find the height h metres of the pyramid a surveyor measures the distance d metres and angles $\alpha°$ ('alpha') and $\beta°$ ('beta').

By investigating triangles ABP and PBM, can you discover a formula involving h, d, α, β that he could use? Use your formula to calculate h, taking $d = 200$, $\alpha = 30$ and $\beta = 20$.

Practical

Use your formula to calculate the height of a local landmark.

THE COSINE RULE

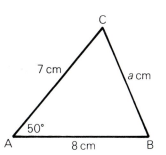

You *could* find a and b by means of scale drawings.
What about trigonometry?
Any right–angled triangles?
The Sine Rule? No?
Back to *making* right–angled triangles . . . and Pythagoras' Theorem!

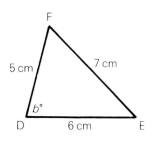

(i) $\angle A$ acute

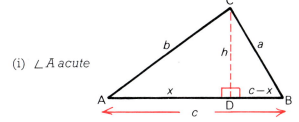

(ii) $\angle A$ obtuse

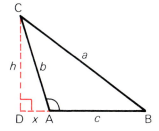

Draw altitude CD to make two right–angled triangles ADC and BDC.
Let $CD = h$ units, and $AD = x$ units.
Using Pythagoras' Theorem in $\triangle BCD$:

(i) $a^2 = h^2 + (c - x)^2$
$\qquad = h^2 + c^2 + x^2 - 2cx$
$\qquad = (h^2 + x^2) + c^2 - 2cx$
$\qquad = b^2 + c^2 - 2bc \cos A,$

$\left(\text{since } \dfrac{x}{b} = \cos A, \ x = b \cos A \right)$

(ii) $a^2 = h^2 + (c + x)^2$
$\qquad = h^2 + c^2 + x^2 + 2cx$
$\qquad = (h^2 + x^2) + c^2 + 2cx$
$\qquad = b^2 + c^2 - 2bc \cos A$

$\left(\text{since } \dfrac{x}{b} = \cos(180° - A) = -\cos A, \ x = -b \cos A \right).$

> In any $\triangle ABC$, $a^2 = b^2 + c^2 - 2bc \cos A$, the Cosine Rule.

In the same way, $\quad b^2 = c^2 + a^2 - 2ca \cos B$
$\qquad\qquad\qquad\ c^2 = a^2 + b^2 - 2ab \cos C.$

Notice: (i) a and A in $a^2 = b^2 + c^2 - 2bc \cos A$
(ii) the 'cyclic symmetry' of the letters and angles from one formula to the next.

Example In $\triangle ABC$, $b = 3$, $c = 5$ and $\angle A = 35°$. Calculate a.

$a^2 = b^2 + c^2 - 2bc \cos A$
$\quad = 3^2 + 5^2 - 2 \times 3 \times 5 \times \cos 35°$
$\quad = 9·43$
$a = 3·1$, to 1 decimal place.

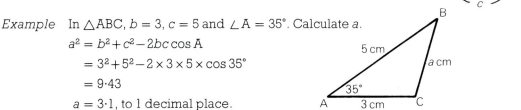

1A

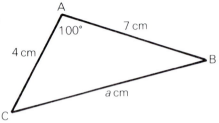

To calculate a, copy and complete:

$$a^2 = b^2 + \ldots - 2bc \cos \ldots$$
$$= 4^2 + \ldots - 2 \times 4 \times 7 \ldots$$
$$= \ldots$$
$$a = \ldots, \text{ to 1 decimal place.}$$

2A Use the Cosine Rule to find a in each of the following:

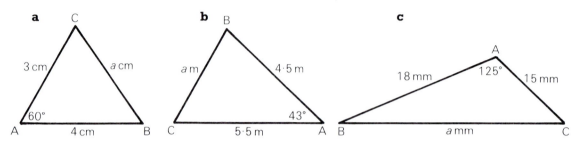

a

b

c

3A a For $\triangle$PQR, write down the Cosine Rule formula for: (i) p^2 (ii) q^2 (iii) r^2.
   b Calculate the length of the third side, given:
   (i) $q = 10, r = 7$ and $\angle$P = 48° (ii) $p = 4\cdot5, q = 5\cdot2, \angle$R = 118°.

4A Clare, an apprentice surveyor, sketches the boating pond in her pad, and marks in these measurements. Calculate the width of the pond (AB).

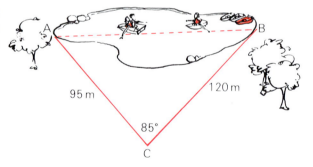

5A

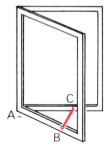

The window is held open by the usual kind of bar, hinged at B and with holes along it. The holes fit over a peg C. Calculate BC when the window is held open at an angle of 40°, AB = 45 cm and AC = 50 cm.

6B For $\triangle$ABC, pair off each of these

$\angle$A = 90°
$\angle$A > 90°
$\angle$A < 90°

with one of these

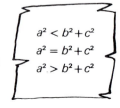

$a^2 < b^2 + c^2$
$a^2 = b^2 + c^2$
$a^2 > b^2 + c^2$

7B Calculate *a* for each wing position in this 'swing-wing' aircraft:

(i) 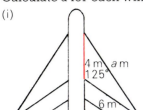 (ii) 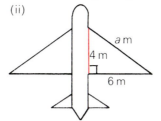 (iii)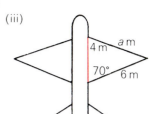

8B Sketch a parallelogram ABCD in which AB = 10 cm, AD = 8 cm and ∠BAD = 60°. Calculate the lengths of its diagonals.

9B The planet Mars (M) has two moons. Phobos (P) is 5900 miles and Deimos (D) is 14 600 miles from the planet's centre, to the nearest 100 miles. How far apart are the moons, to the nearest 100 miles, when:

  a ∠PMD = 70°   **b** ∠PMD = 140°?

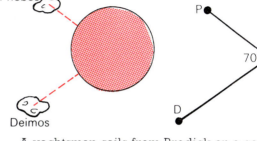

10B

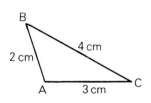

FIRTH OF CLYDE

A yachtsman sails from Brodick on a course bearing 050°. After 15 km, at C, he changes to a bearing of 112°, and sails 11 km to Ardrossan. A car ferry sails directly between Brodick (B) and Ardrossan (A). Calculate:

  a ∠BCA
  b the length of the car ferry's journey
  c ∠CBA
  d the bearing of Ardrossan from Brodick.

Calculating an angle of a triangle, given the lengths of the sides

To find the size of angle A:
Make a scale drawing?
Use trigonometry—Right–angled △?
              —Sine Rule?
              —Cosine Rule?____↗

$$a^2 = b^2 + c^2 - 2bc\cos A.$$

$$2bc\cos A = b^2 + c^2 - a^2$$

$$\cos A = \frac{b^2 + c^2 - a^2}{2bc}.$$

Example   Calculate the largest angle in △ABC above.

The largest angle is opposite the largest side, i.e. ∠A.

$$\cos A = \frac{b^2 + c^2 - a^2}{2bc}$$

$$= \frac{3^2 + 2^2 - 4^2}{2 \times 3 \times 2}$$

$$= -0.25$$

$$\angle A = 104.5°.$$

1A Write down formulae for cos A, cos B and cos C in terms of a, b and c. (Remember the 'cyclic symmetry' $a \to b \to c \to a \ldots$.)

2A In $\triangle$ABC, calculate:

a $\angle$ A when $a = 4$, $b = 5$, $c = 6$ **b** $\angle$ B when $a = 31$, $b = 42$, $c = 53$

c $\angle$ C when $a = 2 \cdot 5$, $b = 4 \cdot 5$, $c = 3 \cdot 5$.

3A In $\triangle$PQR, $p = 8$, $q = 12$ and $r = 10$.

a Write down formulae for cos P, cos Q and cos R.

b Calculate the size of: (i) the largest angle   (ii) the smallest angle, in the triangle.

4A Use the Cosine Rule to calculate the sizes of the smallest and largest angles of $\triangle$XYZ, in which $x = 105$, $y = 125$ and $z = 205$.

5A

Stan's ladder has legs 150 cm and 146 cm long. When the ladder is fully open, the feet are 86 cm apart.

a Calculate the angles of the triangle formed when the ladder is fully open.

b In a narrow space, Stan can only get the feet 80 cm apart. Calculate the angle between the legs of the ladder then.

6B Aimee is measuring the floor of her lounge for a fitted carpet. Calculate:

a the sizes of the angles at the corners of the room

b the length of the other diagonal.

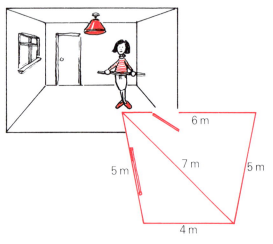

7B

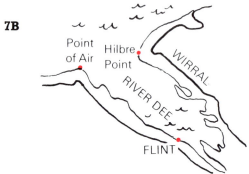

Hilbre Point is 9 km from Point of Air, and 15·2 km from Flint. Point of Air is 17·8 km from Flint.

If Hilbre Point bears 070° from Point of Air, find the bearing of Flint from Point of Air.

Eric and Clare, surveyors from Crookes and Angles, have to find the distance between two far–off landmarks at C and F. They measure the length of a baseline AB, and the sizes of the marked angles.
By investigating triangles ABC, ABF and AFC can you discover a formula involving d, u, v, w and x for calculating the distance CF?

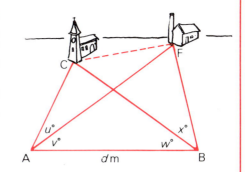

Use your formula to calculate CF when $d = 200$, $u = 10$, $v = 40$, $w = 50$ and $x = 15$.

Practical

Use your formula to calculate the distance between two local landmarks.

THE AREA OF A TRIANGLE

Step by step

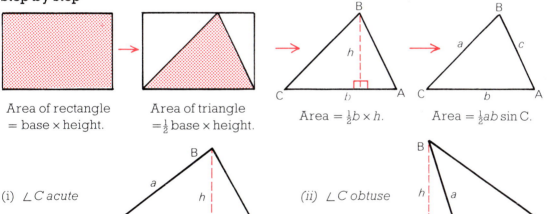

Area of rectangle = base × height.

Area of triangle = $\frac{1}{2}$ base × height.

Area = $\frac{1}{2}b \times h$.

Area = $\frac{1}{2}ab \sin C$.

(i) $\angle C$ acute

(ii) $\angle C$ obtuse

Draw altitude BD to make two right–angled triangles BDC and BDA. Let BD = h units.

The area of $\triangle ABC = \frac{1}{2}bh$

$$= \frac{1}{2}b \times a \sin C \left(\text{In (ii)}, \frac{h}{a} = \sin(180° - C) = \sin C \right)$$

$$= \frac{1}{2}ab \sin C.$$

The area of $\triangle ABC = \frac{1}{2}ab \sin C.$

In the same way, the area of $\triangle ABC = \frac{1}{2}bc \sin A = \frac{1}{2}ca \sin B.$

Example

Calculate the area of △ABC with $a = 3$, $b = 6$ and $\angle C = 70°$. The lengths are in centimetres.

$$\text{Area} = \tfrac{1}{2}ab\sin C$$
$$= \tfrac{1}{2} \times 3 \times 6 \times \sin 70°$$
$$= 8{\cdot}5, \text{ to 1 decimal place.}$$

The area is $8{\cdot}5\,\text{cm}^2$.

===== *Exercise 5* =====

1A Calculate the area of each triangle:

a   **b**   **c**

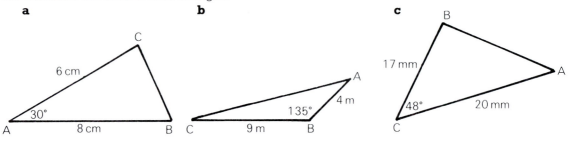

2A Calculate the area of △ABC, with:
a $\angle A = 100°$, $b = 2{\cdot}7$, and $c = 4$. The lengths are in centimetres.
b $a = 7{\cdot}1$, $b = 3{\cdot}5$ and $\angle C = 21{\cdot}5°$. The lengths are in metres.

3A Calculate the area of △XYZ, in which $x = 44$, $z = 55$ and $\angle XYZ = 66°$. The lengths are in millimetres.

4A Calculate the area of parallelogram PQRS. **5A** Calculate the area of farmer Jones' field.

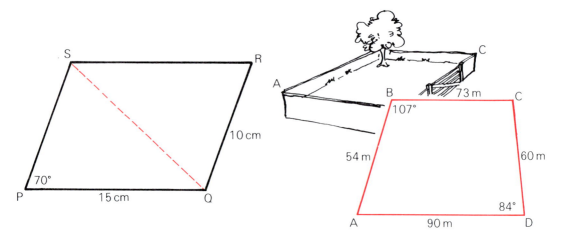

6B The area of △ABC is $12\,\text{cm}^2$. AC $= 5\,\text{cm}$ and BC $= 6\,\text{cm}$. Calculate two possible sizes of $\angle$ C.

7B The body of Ron's barrow is made from a sheet of aluminium. The front, base and back are rectangles. The two sides are congruent quadrilaterals. Calculate the area of aluminium sheet needed for the barrow. The lengths are in centimetres.

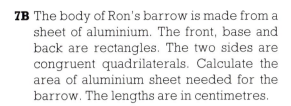

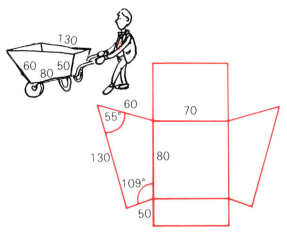

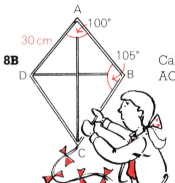

8B Calculate BC, and then find the area of Astrid's kite. Remember that AC is an axis of symmetry.

WHICH FORMULA?

=== *Exercise 6* ===

1A Calculate AC.

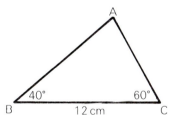

2A Calculate RQ.

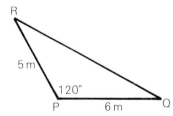

3A Calculate ∠UVW.

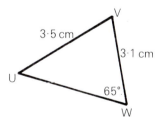

4A Sketch a triangle with sides 5 cm, 6 cm and 10 cm long. Calculate the sizes of all the angles in the triangle.

5A Calculate: **a** BC **b** the area of △ABC.

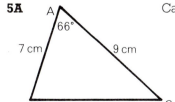

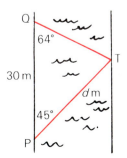

6A P and Q are points on a river bank 30 m apart. T is a tree on the opposite bank. Calculate: **a** d **b** the width of the river. (Copy the triangle, and draw TR perpendicular to PQ.)

7A Calculate:
  a the total length of the cord (P to Q to R to P)
  b ∠ QPR when the picture is straightened.

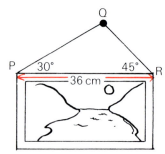

8B A farmer uses two sections of fencing against a wall to enclose an area of 26 m². One section is 12 m long and the other is 9 m long.
Calculate: **a** two possible values of angle ABC
  b the corresponding lengths of AC.

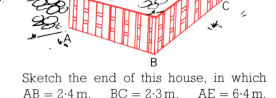

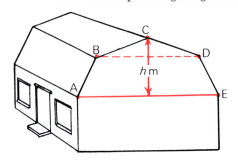

9B

Sketch the end of this house, in which AB = 2·4 m,   BC = 2·3 m,   AE = 6·4 m, ∠ EAB = 63° and ∠ DBC = 20°. Calculate:
a ∠ ABC (Extend AB, if necessary.)
b AC   **c** the height *h* m of the roof space.

Every regular polygon can be divided into congruent isosceles triangles. For example:

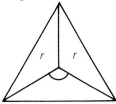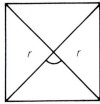

a For each one, calculate:
   (i) the size of an angle at the centre
   (ii) the area of one triangle, in terms of *r*
   (iii) the area of the polygon in terms of *r*.
b Repeat the steps above to find a formula for the area of an *n*–sided polygon, in terms of *n* and *r*.
c (i) As *n* increases (more and more sides) the polygon's shape approaches a circle. Use the approximation 'Area of circle ≑ area of polygon' to prove that
   $\dfrac{n}{2}\sin\left(\dfrac{360°}{n}\right)$ is a good approximation for π.
   (ii) What approximation does your calculator give for π?
d Using your calculator, investigate how many sides a polygon must have before it gives as good an approximation as this.

e Write a computer program which will print out successive approximations for π as *n* increases.

TRIANGLES AND TRIGONOMETRY

a (i) Construct △ABC with BC = 7 cm, BA = 6 cm and CA = 5 cm.
   (ii) Construct the lines bisecting the angles at B and C, to meet at I.
   (iii) With centre I, and radius IK, draw the 'inscribed' circle in △ABC. Measure IK.

b Calculate the length of IK. You'll need the Cosine Rule, the Sine Rule and a 'trig' ratio' in a right-angled triangle.

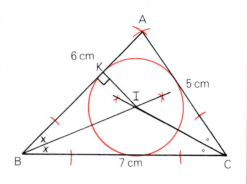

CHECK-UP ON **TRIANGLES AND TRIGONOMETRY**

1 Copy and complete the following for △ABC:

a $\dfrac{a}{\sin A} = \dfrac{\ldots}{\sin B} = \dfrac{c}{\ldots}$ **b** $a^2 = b^2 + \ldots\ldots\ldots$ (Cosine Rule)

c $\cos B = \dfrac{\ldots\ldots - b^2}{\ldots}$ **d** Area of $\triangle = \frac{1}{2}\ldots\ldots \sin A$.

2 Sine Rule, Cosine Rule, Pythagoras' Theorem—which would you use to calculate x in each triangle:

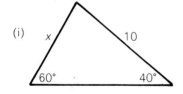
(i)

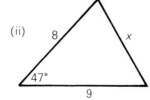
(ii)

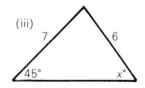
(iii)

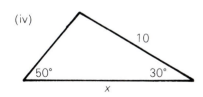
(iv)

(v)

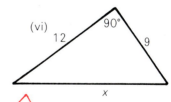
(vi)

3 Calculate x in triangles (i), (ii) and (v) in question **2**.

4 A TV detector van picks up strong signals at A and B.
Calculate the distance from each position to the TV set.

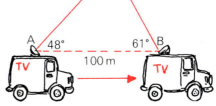

5

B

4 m   5 m

A   C

A loop of rope 16 m long is pegged out to form a triangle. Calculate:

a $\angle B$ **b** the area of $\triangle ABC$.

6 A ship radios its position A when it is 14 km from port P. It radios again at B, after sailing 16 km. At P, $\angle APB = 33°$. Calculate: **a** $\angle B$ **b** PB.

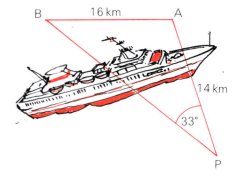

B   16 km   A

14 km

33°

P

7 A plane flies for 100 km on a bearing of 048°. It then flies for 60 km on a bearing of 159°. Make a sketch, and calculate the bearing and distance of its course back to base.

Class discussion

It will probably stay fair today.

It is likely to be pie and chips for lunch.

Rovers have a 50-50 chance of winning.

The odds are against people landing on Mars before the year 2000.

'silver sails' has a good chance of winning the cup.

success, or failure, in choosing a red card from a pack are equally likely.

Can you think of other situations in everyday life where 'probable', 'equally likely', 'odds', 'chance' are used?

All of these have a common theme—chance, or probability.

Why is a game often started by—

Tossing a coin?

or

Rolling a dice?

or

Drawing a card?

What other ways can you think of?

RELATIVE FREQUENCY AND PROBABILITY

=== *Exercise 1* ===

1 *A problem. 'What fraction of the population is left–handed?'*

Is it 0·1, or 0·5, or 0·7, or are people equally likely to be left or right–handed? Have a guess. Now find out.

a Copy and complete this table for your own class:

	Number	Fraction of total
Left–handed		
Right–handed		
TOTAL		

Correct to 2 decimal places

153

b Calculate the **Relative Frequency** of left–handed pupils in your class, that is

$$\frac{\text{Number who are left–handed}}{\text{Total number in class}}.$$

c Here are tables of larger samples of people. Calculate the relative frequency of left–handed people for each one.

(i)

Left–handed	22
Right–handed	78
TOTAL	
Relative frequency	

(ii)

Left–handed	54
Right–handed	146
TOTAL	
Relative frequency	

(iii)

Left–handed	248
Right–handed	752
TOTAL	
Relative frequency	

d Draw a bar-graph of the four results on squared paper, using the scales shown. The larger the sample, the more likely is the relative frequency to be closer to 0·25. This number is called the **Probability** that a person chosen at random is left–handed.

 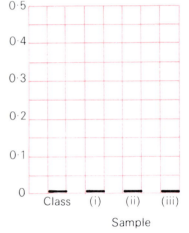

$$\boxed{Experiment \rightarrow \text{calculation of } relative\ frequency \rightarrow probability.}$$

2 *Another problem that can be solved by collecting and analysing data.*
'What is the probability of success with this new medical treatment?'

	Year 1	Years 1 and 2	Years 1–3	Years 1–4	Years 1–5
Number of successes	7	24	46	46	80
Number of patients	10	30	50	60	100
Relative frequency					

a Calculate the relative frequencies, then estimate the probability of success.
b Draw a bar-graph of the results of the trials.

More experiments

Try some of the experiments below. For each one:

a Guess the probability.

b Do the experiment 100 times, recording your results in a tally table.

c Calculate the fraction of successful outcomes ('Pin-up' in I, a '6' in II, and so on).

d Estimate the probability of the outcome. How good were your guesses?

I. *Probability of 'pin up' when a drawing pin is dropped onto a flat surface.*

II. *Probability of a '6' when a dice is rolled.*

III. *Probability of a 'Head' when a coin is tossed.*

IV. *Probability of cutting an 'Ace' from a pack of cards; and again after all the Kings, Queens and Jacks have been removed from the pack. Are the probabilities different?*

EQUALLY LIKELY OUTCOMES

In a large number of tosses of a coin the fraction of 'Heads' is about 0·5. It's the same for 'Tails'.

When a coin is tossed there are two **equally likely** results, or *outcomes*, a 'Head' or a 'Tail'. The probability of 'Head' is $\frac{1}{2}$. We write P(Head) $= \frac{1}{2}$, or 0·5.

> Where there are several *equally likely* outcomes of an event, the *probability* of a 'favourable' outcome $= \dfrac{\text{number of favourable outcomes}}{\text{number of possible outcomes}}$.

Example

Ten cards are numbered from 1 to 10. The cards are shuffled, and one is drawn from the pack at random. Calculate the probability that: **a** it is the 7 **b** it is an even number.

a Number of favourable outcomes = 1
Number of possible outcomes = 10 (all equally likely)
$$P(7) = \frac{1}{10}, \text{ or } 0·1, \text{ or } 10\%.$$

b Number of favourable outcomes = 5
Number of possible outcomes = 10 (all equally likely)
$$P(\text{even number}) = \frac{5}{10} = \frac{1}{2}, \text{ or } 0·5, \text{ or } 50\%.$$

INTRODUCING PROBABILITY

1A List the possible outcomes on rolling a dice. What is the probability of scoring only 1?

2A A pencil–case holds red, blue, green, black and yellow crayons—one of each. Shona chooses a crayon at random.
  a How many possible outcomes are there?
  b Calculate the probability that she chooses the red one.

3A Ian's game needs a special pack of cards with all the letters of the alphabet. Each card has a different letter on it. Tom draws a card.
  a How many possible outcomes are there?
  b What is the probability that he draws:
      (i) Z   (ii) a vowel
      (iii) a letter of the word IN?

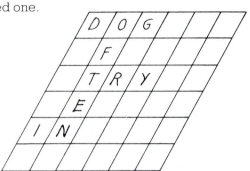

4A

Alison rolls a dice.
  a How many possible outcomes are there?
  b Calculate the probability that she scores:
      (i) an even number   (ii) more than 4.

5A Terry spins a regular octagonal spinner.
  a Calculate the probability that he scores:
      (i) 1   (ii) an odd number   (iii) a multiple of 3.
  b What have you assumed about the spinner?

6A A letter is chosen at random from the word MATHEMATICS.
  Calculate: **a** P(M)   **b** P(a vowel)   **c** P(a consonant).

7A A bag holds 5 red beads and 15 black beads. Stuart takes one out at random.
  a Calculate:
      (i) P(red bead)   (ii) P(black bead).
  b What is the sum of these probabilities?

8A Jasper rolls a dice. He needs 3 or more to win the game.
  a List the set of all possible outcomes.
  b List the set of favourable outcomes.
  c Calculate the probability that he wins.

9A The bar–graph shows the number of pupils in Miss Jack's class with different colours of hair.

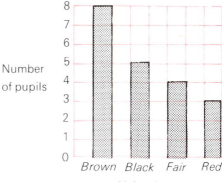

a Calculate the probability that a pupil chosen at random will have:
   (i) red hair   (ii) brown hair.

b How would you choose a pupil at random?

10A Mick and Kay are playing battleships on two 10×10 grids of squares. Each has:

| 1 submarine | 1 frigate | 1 aircraft carrier | 1 battleship |

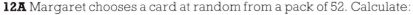

Mick has first shot. He picks a square at random. Calculate the probability that he hits:
a the submarine   **b** the battleship.

11A

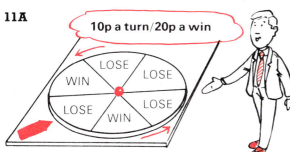

10p a turn/20p a win

LOSE LOSE
WIN
LOSE LOSE
WIN

a Calculate the probability of:
   (i) winning 20p
   (ii) losing 10p.

b Add up the two probabilities.

Have a go! All sectors equally alike.

12A Margaret chooses a card at random from a pack of 52. Calculate:

a P(King of Hearts)   **b** P(A red king)   **c** P(a king)
d P(a heart)   **e** P(A king, queen or jack).

13B

Brian's class is holding a raffle. 100 tickets have been sold. Brian buys two tickets. There are two prizes.

a What is the probability that the first number drawn is one of Brian's?

b Brian wins the first prize. What is the probability that he will win the other prize?

14B In a box of 100 light bulbs, 90 are known to work and the remainder are faulty. The tester chooses a bulb at random.

a Calculate the probability that it works.

b The bulb is faulty, and is destroyed. Calculate the probability that the next bulb chosen will work.

c Would your answers be different for 1000 bulbs where 10% are known to be faulty? Explain.

Mr Adamson has two pigeons, Patti and Pete. He is going to make a new house for them. His first plan has two rooms, one for Patti and one for Pete. He then decides to make three rooms, and let the birds choose whichever ones they wish to live in—one to a room.

a What is the probability that Patti and Pete would be next–door neighbours in each house?

b What would the probability be for 4, 5, 6 . . . , n rooms?

c What would the answers be if the rooms were built in a circle?

CERTAIN SUCCESS AND CERTAIN FAILURE

Sandra has a special two–headed coin.

She tossed it 5, 10, 20 times and wasn't surprised when she got 'Heads' every time, never 'Tails'.

So P(H) = 1, and P(T) = 0. | P(certain success) = 1, and P(certain failure) = 0.

Certain failure   *Certain success*

0   0·5   1

Probability can be measured on a scale from 0 to 1 inclusive.

Exercise 3

1A Draw a scale like the one above. Then, judging with your eye, put arrows where you'd place the probability that:

a a football captain wins the toss of a coin

b an odd number times an odd number makes an odd number

c you will live forever

d on rolling a dice a number less than 5 will turn up

e lightning will strike twice in the same place

f on rolling a dice the number 1 turns up

g $m > n$, if m is a negative number and n is a positive number, and m and n are chosen at random

h you have given the correct answers to these eight questions.

2A A coin is tossed. Calculate: **a** P(Head)   **b** P(Tail)   **c** P(Head) + P(Tail).
Could you have forecast the answer to **c**? What possibility is being ignored?

3A A dice is rolled. Calculate: **a** P(an even number)   **b** P(an odd number)
   c P(an even number) + P(an odd number). Could you have forecast the answer to **c**?

4A A dice shaped like a regular tetrahedron (a pyramid on a triangular base) has numbers 1, 2, 3 and 4 on its sides. The dice is rolled, and the number on the base is noted. Calculate:

a P(4)   **b** P(a number less than 4)   **c** P(4) + P(a number less than 4).

> For a result A, if $P(A) = x$, then $P(\text{not }-A) = 1 - x$, where $0 \leqslant x \leqslant 1$.

5A The probability of a seed germinating in a trial of new flowers is 0·9. What is the probability that it will *not* germinate?

6A It is estimated that in a home game the probability that Rovers will fail to score is 0·43. What is the probability that they will score at least 1 goal?

7B The probability of scoring 7 with two dice is $\frac{1}{6}$.

    a Calculate the probability that the score will not be 7.

    b What is the probability that the score is 6?

---

1 A dice is rolled, and s stands for the score.
    a Calculate: (i) $P(s \geqslant 4)$ (ii) $P(s < 5)$.
    b Why is $P(s \geqslant 4) + P(s < 5) \neq 1$?

2 Robin enters the lift on the fifth floor, and presses one of the buttons at random.
    a Calculate:
      (i) P(going up)   (ii) P(going down).
    b Why is the sum of the probabilities not 1?

Odds. People often give probability in terms of *odds*. Tick–Tack Terry certainly does this at the race course.

Horse	Probability of winning	Probability of not winning	Odds
Surely	$\frac{1}{5}$	$\frac{4}{5}$	4 to 1 against
Speedy	$\frac{3}{4}$	$\frac{1}{4}$	3 to 1 on
Steady	$\frac{1}{2}$	$\frac{1}{2}$	1 to 1, or 'evens'.

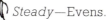

Surely—4 to 1 against   *Speedy*—3 to 1 on   *Steady*—Evens.

    a Do the fractions in the first column add up to 1?
    b Would you expect the sum to be 1, less than 1, or more than 1?
    c Investigate the odds in some of the races shown in newspapers.

FORETELLING THE FUTURE—EXPECTATION

Insurance
How long are you likely to live?

Commerce
How many sales will there be?

Biology
what is the probable yield of wheat?

Space Research
what is the chance of rocket failure?

Games
How often do three lemons appear?
WIN WIN

Politics
which party is likely to win the next general election?

Earthquakes
How many will there be next year?

Example

Smiling Jim spins the TV Wheel of Fortune. Ann watches with bated breath. All stations, they say, are equally likely. What if the pointer hits the jackpot—£100!

$$P(£100) = \frac{1}{6}.$$

For 100 contestants, Jim might expect to pay the jackpot to $\frac{1}{6} \times 100 = 16$ or 17 of them. Can you see why?

WHEEL OF FORTUNE

═══════════ *Exercise 4* ═══════════

1A Look back to the Wheel of Fortune.
   a What is the probability of winning £10?
   b Of 50 contestants, how many are likely to get £10?

2A Approximately how many Heads would you expect in 1000 tosses of a coin?

3A A dice is rolled.
   a Calculate the probability of scoring a prime number.
   b How many of these would you expect in 100 rolls?

4A A dice is rolled 180 times, and a 5 or 6 turns up 120 times. Do you think that the dice is biased? Explain in a sentence.

5A

New Tomato seed—
Red Glory—95% germination

A nurseryman sows 3000 seeds. How many plants could he expect?

JACKPOT £5
WIN WIN

6A The probability of winning the jackpot on the fruit machine is 0·95%. How many jackpots would you expect to win in 250 spins?

Would you agree that in a number of trials, the Expected Frequency of an event = the Probability of the outcome × the number of trials?

7A Some statistics from a maternity unit.
Of 1 000 000 births, estimate the number of sets of:
a twins   **b** triplets   **c** quadruplets.

	Probability
Twins	$\frac{1}{80}$, or 0·0125
Triplets	$\frac{1}{2500}$, or 0·0004
Quadruplets	$\frac{1}{64\,000}$, or 0·000 015 6

8A a A sample of 300 people is chosen at random. How many would you predict were born:
(i) in a leap year   (ii) on 29th February?
b Calculate the relative frequency of leap year birthdays for the pupils in:
(i) your class   (ii) your year group.
Is your answer close to the probability you used in **a**? If not, why not?

9B Dave took a multi–choice test of 20 questions. Each question had four possible answers A, B, C, D.
a Dave did not know his work, so guessed the answers, and got 6 correct. Was he lucky?
b How many would you expect him to get right by chance if there were five answers A, B, C, D, E?

10B

○
○
○ P (accident to driver under 25) = 0·2
○ P (accident to driver 25 or over) = 0·12
    Number of clients
○   (i)   Under 25 — 1200
○   (ii)   25 or over — 3500
○

Memo to salespeople in the Stoutheart Insurance Company. Estimate for the year:
a the total number of accident claims expected
b the number of clients aged 25 or over who are not expected to make a claim.

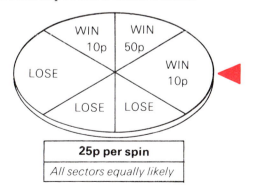

25p per spin
All sectors equally likely

11B a In 300 spins, how often would you expect to win: (i) 50p   (ii) 10p?
b Calculate the expected profit or loss in 300 spins.

Design a 'wheel of fortune' which has eight sectors, costs 25p a spin, and has to make a profit of 10% for the owner.

CHECK-UP ON **INTRODUCING PROBABILITY**

1 A dice is rolled.

Number of throws	20	40	60	80	100
Number of 3s	4	7	11	12	16

   a Calculate the relative frequency, correct to 2 decimal places, of scoring 3 in 20, 40, 60, 80 and 100 throws.
   b Estimate the probability of scoring 3.

2 The numbers in bingo run from 1 to 100. What is the probability that the first number drawn is:
   a 50   **b** an even number   **c** the cube of a whole number?

3 In a survey of 5000 homes, 2600 are found to have central heating. Calculate the probability that a home chosen at random:
   a will have central heating   **b** will not have central heating.

4 A letter is chosen at random from the word TELEVISION. Calculate:
   a P(T)   **b** P(E)   **c** P(Y).

5 Greg's spinner is equally likely to stop at 1, 2, 3, 4 or 5. Calculate the probability of the spinner stopping at:
   a an odd number   **b** an even number
   c a number greater than 1   **d** a number n, where $n \leqslant 1$.

6 Clearview Glazing Co leaflet the town. They estimate that the probability of a response is 3%. From 6500 leaflets, how many enquiries should they expect?

7 Football league games one Saturday resulted in 10 draws, 32 home wins and 12 away wins. A game is chosen at random. Calculate the probability that it will be:
   a a draw   **b** a home win   **c** an away win   **d** a draw, home win or away win.

8

	Second dice					
	1	2	3	4	5	6
First 1	1,1	1,2	1,3	1,4		
dice 2	2,1	2,2				
3	3,1					

Copy and complete the table for the scores when two dice are rolled.
   a How many different totals are possible?
   b Use the entries in the table to calculate the probability of a total of:
       (i) 2   (ii) 3   (iii) 4   (iv) 7.
   c What is the most likely score?

INDICES

Saving space and time

Eight–year old Ewan doesn't know anything about indices, so he needs a whole line for:

$4 \times 4.$

But sixteen–year old Sandra knows all about indices. She saved a lot of space and time by writing 4^{30}. Why 30?

4^{30} is read '4 *to the power* 30'. 30 is called the *index* of the power.

Exercise 1

1A Write, as Ewan would: **a** 4^5 **b** x^3 **c** $2y^2$ **d** $3t^4$ **e** 3×10^4 **f** 2×2^5.

2A Write, as Sandra would:
a $3 \times 3 \times 3 \times 3 \times 3$ **b** $x \times x \times x \times x \times x$ **c** $a \times a \times b \times b \times b$ **d** $10 \times 10 \times 10 \times 10 \times 10 \times 10$.

3A Use your calculator to find 4^{30} like this:

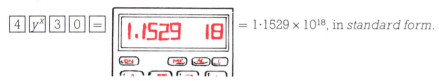 $= 1 \cdot 1529 \times 10^{18}$, in *standard form*.

Why is $1 \cdot 1529 \times 10^{18}$ just an approximation for 4^{30}?

4A Use your calculator to find the exact value of: **a** 2^5 **b** 3^4 **c** 10^6 **d** 10^8.

5A a Use your calculator to find an approximation for: (i) 2^{100} (ii) 20^{20} (iii) 51^{15}.
  b Try 100^{100}. What happens? Why?

6A $p = 3$, $q = 2$ and $r = 1$. You won't need your calculator to find the value of:
  a $p^2 + q^2 + r^2$ **b** $(p + q + r)^2$ **c** $(2p)^2$ **d** $2p^2$ **e** $3q^3$.

7B $x = 2$ and $y = -2$. Find the value of:
  a $x^2 + y^2$ **b** $x^3 + y^3$ **c** $(x + y)^4$ **d** $(x - y)^4$ **e** $3x^2 - 4y$.

8B What is the smallest value of x in 2^x for which your calculator has to use standard form?

Exercise 2

1A a Copy and complete: (i) $2^5 \times 2^2 = 2 \times 2 \times \ldots \ldots = 2^{\cdots}$
                          (ii) $a^3 \times a^2 = a \times a \times \ldots \ldots = a^{\cdots}$
                          (iii) $b^4 \times b^4 = b \times b \times \ldots \ldots = b^{\cdots}$.
  b Can you see a quick way to write down the answers?

To multiply, you add indices

$$a^m \times a^n = a^{m+n}$$

2A Simplify the following: for example, $2a^4 \times 5a^2 = 2 \times 5 \times a^4 \times a^2 = 10a^6$.

   a $x^2 \times x^3$ **b** $y^4 \times y^2$ **c** $d^3 \times d^6$ **d** $t^5 \times t^5$ **e** $k^2 \times k$

   f $2^3 \times 2^7$ **g** $3^6 \times 3^6$ **h** $4^8 \times 4$ **i** $m^{10} \times m^{10}$ **j** $n^2 \times n^2 \times n^2$

   k $2a^2 \times 5a^2$ **l** $3p^2 \times 2p^3$ **m** $x^8 \times 5x^2$ **n** $y^3 \times y^2 \times y$ **o** $z^5 \times z^5 \times z^5$.

3A Break the brackets: for example, $t^2(t^4 + t^7) = t^6 + t^9$

   a $x^3(x^2 + x^4)$ **b** $y^4(y^2 + y^3)$ **c** $a^2(a^3 + a^2 + a)$.

4A Find x: for example, $n^x \times n^2 = n^5$. $n^{x+2} = n^5$.   So $x + 2 = 5$, and $x = 3$.

   a $a^3 \times a^2 = a^x$ **b** $b^x \times b^4 = b^7$ **c** $c^2 \times c^x = c^{2x}$.

5A a Copy and complete: (i) $3^4 \div 3^2 = \dfrac{3^4}{3^2} = \dfrac{3 \times 3 \times \ldots}{3 \times \ldots} = 3^{\cdots}$

                 (ii) $a^6 \div a^3 = \dfrac{a^6}{a^3} = \dfrac{a \times a \times \ldots}{a \times \ldots} = a^{\cdots}$.

  b What is the quick way to write down the answers?

To divide, you subtract indices

$$a^m \div a^n = a^{m-n}.$$

6A Simplify the following: for example, $6a^4 \div 3a^3 = 2a$.

   a $y^6 \div y^4$ **b** $z^3 \div z^2$ **c** $x^{20} \div x^{10}$ **d** $5^4 \div 5^2$ **e** $2^8 \div 2^5$

   f $6a^4 \div 2a^2$ **g** $8b^5 \div 2b$ **h** $3c^6 \div 3c^2$ **i** $2d^2 \div 2d$ **j** $e^{11} \div e^{10}$.

7A Simplify:

   a $\dfrac{2^3 \times 2^8}{2^6}$ **b** $\dfrac{3^4 \times 3^6}{3^5}$ **c** $\dfrac{5^{10}}{5^4 \times 5^4}$ **d** $\dfrac{x^4 \times x^4}{x^3}$ **e** $\dfrac{y^5}{y^2 \times y}$.

8A a Copy and complete: (i) $(5^2)^3 = 5^2 \times 5^2 \times \ldots = 5^{2+2+\cdots} = 5^{\cdots}$

                   (ii) $(a^4)^2 = a^4 \times \ldots = a^{4+\cdots} = a^{\cdots}$.

  b How can you write down the answers this time?

To find a power, you multiply indices

$$(a^m)^n = a^{mn}$$

9A Simplify the following: for example, $(a^4)^2 = a^8$

   a $(y^2)^3$ **b** $(z^3)^5$ **c** $(x^4)^4$ **d** $(t^5)^2$ **e** $(v^2)^{10}$.

10A a Copy and complete: $(ab)^3 = ab \times ab \times ab = \ldots$.

   b Simplify: (i) $(mn)^2$ (ii) $(xy^2)^3$ (iii) $(2k)^2$ (iv) $(pqr)^4$.

11A Make an equation in x, and solve it:

   a $(2^x)^3 = 2^6$ **b** $(3^x)^4 = 3^4$ **c** $(2^x)^2 = 16$ **d** $(5^x)^2 = 25$.

12A Simplify: **a** $x^5 \times x^4$ **b** $y^8 \div y^3$ **c** $(z^4)^5$ **d** $2^3 \times 2^2$ **e** $3^9 \div 3^8$

           f $(2^3)^2$ **g** $3a^2 \times 2a^3$ **h** $8b^7 \div 2b^3$ **i** $(cd^2)^5$ **j** $\dfrac{a^3 \times a^4}{a^2 \times a}$.

13B a $5^{2n} = 5^3 \times 5^7$. Find n.   **b** Express $\dfrac{3^3 \times 3^8}{3^6}$ as a power of 3.

14B Solve: **a** $2^x \times 2^{x+1} = 8$ **b** $(2^x)^{x-1} = 4$.

====================== *Exercise 3* ======================

1A a Copy and complete: (i) $x^5 \div x^5 = x^{5-\cdots} = x^{\cdots}$

(ii) $x^5 \div x^5 = \dfrac{x^5}{x^5} = \ldots$

b What can you deduce?

2A a Follow these instructions on your calculator:

(i) $\boxed{2}\ \boxed{y^x}\ \boxed{0}\ \boxed{=}$ (ii) $\boxed{3}\ \boxed{y^x}\ \boxed{0}\ \boxed{=}$ (iii) $\boxed{99}\ \boxed{y^x}\ \boxed{0}\ \boxed{=}$

b What do you find?

$$\boxed{a^0 = 1}$$

3A a Copy and complete: (i) $x^3 \div x^5 = x^{3-\cdots} = x^{\cdots}$

(ii) $x^3 \div x^5 = \dfrac{x^3}{x^5} = \dfrac{x \times x \times x}{x \times \ldots} = \dfrac{1}{x^{\cdots}}$.

b What can you deduce?

4A a Key in: $\boxed{2}\ \boxed{y^x}\ \boxed{1}\ \boxed{+/_}\ \boxed{=}$. Copy and complete: $2^{-1} = \ldots = \dfrac{1}{2}$

b Key in: $\boxed{2}\ \boxed{y^x}\ \boxed{2}\ \boxed{+/_}\ \boxed{=}$. Copy and complete: $2^{-2} = \ldots = \dfrac{1}{2^2}$

c Key in: $\boxed{10}\ \boxed{y^x}\ \boxed{3}\ \boxed{+/_}\ \boxed{=}$. Copy and complete: $10^{-3} = \ldots = \dfrac{1}{10^3}$.

$$\boxed{a^{-m} = \dfrac{1}{a^m}}$$

5A Write with positive indices: for example, $t^{-5} = \dfrac{1}{t^5}$.

a a^{-2} **b** b^{-4} **c** c^{-10} **d** 2^{-3} **e** 5^{-1} **f** x^{-5}.

6A Write down the value of:
a a^0 **b** 10^0 **c** 3^{-2} **d** 2^{-1} **e** 2^0 **f** 10^{-2}.

7A Which is larger?
a 3^2 or 2^3 **b** 5^1 or 1^5 **c** 2^{-3} or 3^{-2} **d** 2^0 or 2^{-1}.

8A Write each as a power of x: for example   (i) $\dfrac{1}{2x} = \dfrac{1}{2}x^{-1}$ (ii) $\dfrac{2}{3x^7} = \dfrac{2}{3}x^{-7}$

    a $\dfrac{1}{x^2}$ **b** $\dfrac{1}{x^4}$ **c** $\dfrac{2}{x^3}$ **d** $\dfrac{1}{3x}$ **e** $\dfrac{2}{x^2}$ **f** $\dfrac{5}{2x^3}$.

9A Express in standard form, $a \times 10^n$ where $1 \leqslant a < 10$ and n is an integer:
    a $3{\cdot}14 \times 10^5 \times 10^3$ **b** $5{\cdot}6 \times 10^4 \times 10^{-8}$ **c** $123 \times 10^{10} \times 10^{-3}$.

10A Simplify:
    a $p^4 \times p^{-3}$ **b** $q^{-2} \times q^{-2}$ **c** $r^4 \times r^{-4}$ **d** $u^5 \div u^2$ **e** $v^3 \div v^{-1}$
    f $w^{-3} \div w^{-1}$ **g** $(x^2)^{-3}$ **h** $(y^{-1})^{-1}$ **i** $(z^{-4})^0$ **j** $(k^0)^{-2}$.

11B Find the value of: **a** $(2^3)^{-1}$ **b** $(3^{-2})^{-1}$ **c** $(\tfrac{1}{2})^{-2}$ **d** $(\tfrac{1}{3})^{-3}$ **e** $(\tfrac{3}{4})^{-1}$.

12B Break brackets, and finish with positive indices: $x^{-3}(x^4 - x^2) = x - x^{-1} = x - \dfrac{1}{x}$

    a $y^4(y^2 + y^{-2})$ **b** $k^3(k + k^{-1})$ **c** $n^{-2}(n^3 + n)$ **d** $m^{-1}(m + m^2)$
    e $u(u^2 - u^{-2})$ **f** $v(v^{-1} - v^{-2})$ **g** $w^{-3}(2 + w^{-1})$ **h** $z^{-1}(z^{-1} - 2z)$.

13B A rectangular region of sea surface is $8{\cdot}5 \times 10^5$ m long and $4{\cdot}5 \times 10^3$ m broad. Calculate its area, using standard form.

14B The mass of an electron is $9{\cdot}11 \times 10^{-28}$ g. Find an estimate in standard form for:
    a the mass of 6×10^{17} electrons.
    b the number of electrons that have a mass of $1{\cdot}09 \times 10^{48}$ g.

Start with $a^m \times a^n = a^{m+n}$.

    a Put $n = 0$, and prove that $a^0 = 1$.

    b Put $n = -m$, and prove that $a^{-m} = \dfrac{1}{a^m}$.

    c Put $m = n = \tfrac{1}{2}$, and try to find a meaning for $a^{1/2}$.

Note   So far you have used the 'laws of indices' for integers m and n. We assume that the rules remain true for all rational numbers.

The mystery of $a^{\frac{1}{3}}$

I Know what a^2, a^{100}, a^0, a^{-1}, a^{-10} mean, but what does $a^{\frac{1}{3}}$ mean?

Sandra was certainly right to be puzzled! But the answer is not hard to find:

$a^{1/3} \times a^{1/3} \times a^{1/3} = a^{1/3 + 1/3 + 1/3} = a$ $a^{2/3} = (a^{1/3})^2 = (\sqrt[3]{a})^2$, **or**

So $(a^{1/3})^3 = a$ $a^{2/3} = (a^2)^{1/3} = \sqrt[3]{a^2}$

    $a^{1/3} = \sqrt[3]{a}$, taking the cube root of each side.

$$a^{m/n} = (\sqrt[n]{a})^m = \sqrt[n]{a^m}.$$

Examples

1 $16^{1/2} = (2^4)^{1/2} = 2^2 = 4$, or $16^{1/2} = \sqrt{16} = 4$

2 $8^{-2/3} = (2^3)^{-2/3} = 2^{-2} = \dfrac{1}{4}$, or $8^{-2/3} = \dfrac{1}{8^{2/3}} = \dfrac{1}{(\sqrt[3]{8})^2} = \dfrac{1}{2^2} = \dfrac{1}{4}$.

Exercise 4

1A Find the values of these (without using your calculator):
   a $9^{1/2}$ **b** $8^{1/3}$ **c** $16^{1/4}$ **d** $100^{1/2}$ **e** $27^{1/3}$ **f** $81^{1/2}$
   g $4^{-1/2}$ **h** $25^{-1/2}$ **i** $8^{-1/3}$ **j** $16^{-1/2}$ **k** $1000^{1/3}$ **l** $1^{1/2}$.

2A Find the value of:
   a $8^{2/3}$ **b** $27^{2/3}$ **c** $64^{1/2}$ **d** $49^{-1/2}$ **e** $16^{-3/2}$ **f** $(-8)^{1/3}$.

3A Find the value of:
   a $2x^{-1/3}$ when $x = 8$ **b** $3y^{-2/3}$ when $y = 27$ **c** $5z^{-3/2}$ when $z = 25$.

4A Given $f(x) = 4^x$, calculate: **a** $f(\tfrac{1}{2})$ **b** $f(-\tfrac{1}{2})$ **c** $f(\tfrac{3}{2})$ **d** $f(0)$.

5A Which is easier to calculate: $(125^{1/3})^2$ or $(125^2)^{1/3}$? What is its value?

6A Given $x = 36$, $y = 27$ and $z = 16$, find the value of:
   a $x^{1/2} + y^{1/3} + z^{1/4}$ **b** $y^{4/3} \times z^{5/4} \div x^{3/2}$.

7A a Find the value of: (i) $100^{3/2}$ (ii) $81^{3/4}$.
   b Check your answers on your calculator for: (i) $100^{1.5}$ (ii) $81^{0.75}$.

8A Calculate the length of the side of a cube which has a volume of 1728 cm³.

9B Simplify:
   a $x^{1/2}(x^{1/2} + x^{-1/2})$ **b** $y^{2/3}(y^{4/3} - y^{1/3})$ **c** $z^{-1/2}(z^{3/2} + z^{-1/2})$.

10B Simplify:
   a $a^{1/2} \times a^{3/2}$ **b** $b^{3/4} \div b^{1/4}$ **c** $c^{-2/3} \times c^{5/3}$ **d** $d^{-1/2} \div d^{-3/2}$
   e $(u^{1/2})^2$ **f** $(v^{-1/3})^3$ **g** $3n^{1/4} \times 4n^{5/4}$ **h** $8m^{4/5} \div 2m^{3/5}$.

11B Express as a sum of powers of t: for example, $\dfrac{t^{2/3} + 2}{t^{1/3}} = t^{1/3} + 2t^{-1/3}$

   a $\dfrac{t+1}{t^{1/2}}$ **b** $\dfrac{t^{3/2} + t}{t^{1/2}}$ **c** $\dfrac{2t-1}{t^{1/3}}$ **d** $\dfrac{4t^2 - 3t}{2t^{1/3}}$.

Astronomical distances are measured in 'astronomical units', 'light years' and 'parsecs'. Investigate the meaning and magnitude of these units (in kilometres, in standard form). Calculate the relations between the units.

GROWTH AND DECAY

Growth and decay are all around you—in plants, in populations, in life itself. Sometimes they can be described mathematically. For example:

1. Leave £1 in the bank at 10% p.a. compound interest.
   After 1 year you'll have £1 + £0·1 = £1·1
   After 2 years you'll have £1·1 + £0·11 = £1·21 = £1·1^2
   After 3 years, you'll have £1·21 + £0·121 = £1·331 = £1·1^3
   The formula for the amount £A after n years is $A = 1·1^n$.

2. A profit of £1 is doubled every year in future. *The formula for the profit £P after n years is* $P = 2^n$.

3. A loss of £1 is halved every year in future.
   The formula for the loss £P after n years is $P = \dfrac{1}{2^n} = 2^{-n}$.

4. Unit mass of radioactive mineral decays after t years according to
   the formula $m = e^{-0·05t}$, *where* $e \doteqdot 2·718$.

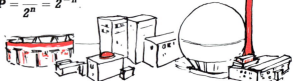

> These four formulae describe *exponential* growth or decay, which have a characteristic equation $y = a^x$ and a graph like the one shown below when $a > 1$.

Exercise 5

1A *The graphs of* $y = 2^x$ *and* $y = 2^{-x}$.

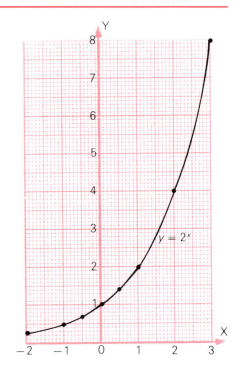

a Copy and complete the table:

x	-2	-1	0	1	2	3
2^x	0·25					
2^{-x}	4					

b Draw both graphs on the same sheet.
c (i) Which line is an axis of symmetry of the diagram?
 (ii) Calculate 2^{23} and 2^{-23}.
 (iii) Which line is an asymptote to both curves?
d Which of the curves illustrates:
 (i) growth (ii) decay?

2A Draw the graphs of $y = 3^x$ and $y = 3^{-x}$ for $x = -3$ to 3.

3A a Draw the graph of the compound interest equation $A = 1\cdot1^n$ for $n = 0, 10, 20, 30$ and 40.
    Scales: 10 units to 2 cm on each axis.
   b From your graph, estimate the amount after: (i) 25 years   (ii) 35 years.
    Check your answers by calculator.

4B The mass—time equation for radioactive mineral decay was $m = e^{-0\cdot05t}$.
   a Calculate m for $e = 2\cdot718$ and $t = 0, 10, 20$ and 40.
   b *Sketch* the graph.
   c 'Half–life' is the time for half the mass to remain. Check that this is about 14 years.

Newton's law of cooling

Heat a beaker or pan of water, not to boiling point. Put a thermometer in the water, and measure its temperature every minute until it is nearly constant. Put your readings in a table, and plot the points. Join them by a 'best–fitting' smooth curve.

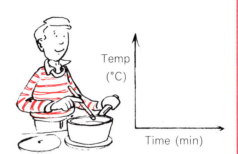

The equation of the curve is of the form $T = a^{-kt}$, where T is the temperature difference between the water and the room, a and k are positive constants and t is the time in minutes from the start.

Sketch the graphs of $y = 2^x$ and $y = 2x + 2$ on the same diagram for $-1 \leqslant x \leqslant 3$. State one value of x for which $2^x = 2x + 2$, and find the other root of this equation, correct to 2 decimal places, using the step-by-step method of approximation.

SURDS

Class discussion

First you solved equations like $2x - 10 = 4$. Solution? (A whole number.)
Then you met equations like $4x + 12 = 4$. Solution? (A negative number.)
And then there was $6x + 5 = 8$. Solution? (A fraction.)

Numbers like these, $7, -2, \frac{1}{2}$, which can be written in the form $\dfrac{p}{q}$, where p and q are integers $(q \neq 0)$ are called *rational numbers*.

But what about the equation $2x^2 + 1 = 5$? Solution? $2x^2 = 4$

$$x^2 = 2$$
$$x = \pm\sqrt{2}$$

... $\sqrt{2}$ is different. It cannot be written $\dfrac{p}{q}$. It is an *irrational number*.

The rational numbers and the irrational numbers
(π is another of these) make up the *real numbers*.
Numbers like $\sqrt{2}$, $\sqrt{3}$, $\sqrt{8}$ and $\sqrt[3]{6}$ are called *surds*.

Exercise 6

1A Which of these are surds? (They cannot 'work out exactly'.)

    a $\sqrt{5}$ **b** $\sqrt{4}$ **c** $\sqrt{6}$ **d** $\sqrt{9}$ **e** $\sqrt{12}$
    f $\sqrt{100}$ **g** $\sqrt[3]{4}$ **h** $\sqrt[3]{8}$ **i** $\sqrt[3]{1000}$ **j** $\sqrt{20}$.

2A Use your calculator for these:

    a $\sqrt{2} \times \sqrt{8}$ ($\boxed{2}$ $\boxed{\text{2ndF}}$ $\boxed{x^2}$ $\boxed{\times}$ $\boxed{8}$ $\boxed{\text{2ndF}}$ $\boxed{x^2}$ $\boxed{=}$) .

    b $\sqrt{2} \times \sqrt{18}$ **c** $\sqrt{5} \times \sqrt{20}$ **d** $\sqrt{6} \times \sqrt{24}$.

Do the answers surprise you? Can you explain them? Try to see how each one works out.

$$\sqrt{a}\sqrt{b} = \sqrt{ab} \quad \text{and} \quad \frac{\sqrt{a}}{\sqrt{b}} = \sqrt{\frac{a}{b}}\,.$$

3A Simplify the following: for example, $\sqrt{75} = \sqrt{(3 \times 25)} = \sqrt{3}\sqrt{25} = 5\sqrt{3}$.
    The aim is to use factors, like 25, which are perfect squares.

    a $\sqrt{12}$ **b** $\sqrt{18}$ **c** $\sqrt{50}$ **d** $\sqrt{28}$ **e** $\sqrt{72}$
    f $\sqrt{200}$ **g** $\sqrt{45}$ **h** $\sqrt{24}$ **i** $\sqrt{96}$ **j** $\sqrt{288}$.

4A Simplify the following: for example, $\sqrt{\dfrac{4}{9}} = \dfrac{\sqrt{4}}{\sqrt{9}} = \dfrac{2}{3}$, or $\sqrt{\dfrac{4}{9}} = \sqrt{\left(\dfrac{2}{3}\right)^2} = \dfrac{2}{3}$

    a $\sqrt{\dfrac{9}{16}}$ **b** $\sqrt{\dfrac{64}{25}}$ **c** $\sqrt{\dfrac{81}{100}}$ **d** $\sqrt{\dfrac{1}{144}}$ **e** $\sqrt{\dfrac{81}{49}}$.

5A Simplify the following: for example, $\sqrt{2} \times \sqrt{6} = \sqrt{2} \times \sqrt{2} \times \sqrt{3} = (\sqrt{2})^2 \times \sqrt{3} = 2\sqrt{3}$
    a $\sqrt{2} \times \sqrt{2}$ **b** $\sqrt{6} \times \sqrt{6}$ **c** $\sqrt{2} \times \sqrt{50}$ **d** $\sqrt{3} \times \sqrt{12}$ **e** $\sqrt{3} \times \sqrt{27}$
    f $\sqrt{2} \times \sqrt{10}$ **g** $\sqrt{3} \times \sqrt{15}$ **h** $\sqrt{5} \times \sqrt{10}$ **i** $\sqrt{5} \times \sqrt{15}$ **j** $\sqrt{6} \times \sqrt{8}$

6A Simplify the following: for example, $6\sqrt{2} - 5\sqrt{2} = \sqrt{2}$.
    a $4\sqrt{2} + 3\sqrt{2}$ **b** $5\sqrt{2} - 2\sqrt{2}$ **c** $7\sqrt{3} + 2\sqrt{3}$ **d** $\sqrt{5} - 4\sqrt{5}$
    e $2\sqrt{10} - 2\sqrt{10}$ **f** $\sqrt{2} + \sqrt{2} + \sqrt{2}$ **g** $\sqrt{3} - \sqrt{3} + \sqrt{3}$ **h** $2\sqrt{7} - 3\sqrt{7} + 2\sqrt{7}$.

7A Simplify:
    a $\sqrt{8} + \sqrt{2}$ **b** $\sqrt{18} + \sqrt{2}$ **c** $4\sqrt{3} - \sqrt{12}$ **d** $\sqrt{45} - \sqrt{20}$.

8A Multiply out, then simplify:
    a $\sqrt{2}(3 + \sqrt{2})$ **b** $\sqrt{3}(\sqrt{2} + \sqrt{3})$ **c** $\sqrt{5}(1 + \sqrt{10})$.

9A $\sqrt{6} = 2 \cdot 45$, to 2 decimal places. Write down approximations for:
    a $\sqrt{600}$ **b** $\sqrt{60\,000}$.

10A A mixture to simplify:

    a $\sqrt{100}$ **b** $\sqrt{300}$ **c** $3\sqrt{6}+2\sqrt{6}$ **d** $3\sqrt{6}\times2\sqrt{6}$

    e $\sqrt{80}+2\sqrt{20}$ **f** $\sqrt{6}\times\sqrt{12}$ **g** $\sqrt{5}(\sqrt{5}+\sqrt{20})$ **h** $2\sqrt{5}-5\sqrt{5}+\sqrt{5}$

    i $x+y$, $x-y$ and xy, given $x=\sqrt{3}+1$ and $y=\sqrt{3}-1$.

11A a Calculate the lengths of AC and AD as
    surds in their simplest form.

    b Write down the length of the hypot-
    enuse of the nth right–angled triangle
    in the sequence $\triangle$ABC, $\triangle$ACD, ...

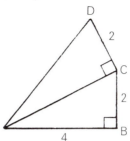

12B The surface area of a cube is $120\,\text{cm}^2$. Calculate the length of its side as a surd.

13B

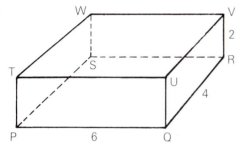

Calculate, as surds in their simplest form,
the lengths of:

a the face diagonals

b the space diagonals, of this cuboid.

A trick of the trade — rationalising a denominator

Use your calculator for: **a** $\dfrac{1}{\sqrt{2}}$ **b** $\dfrac{\sqrt{2}}{2}$.

Same answer? Why?

$$\frac{1}{\sqrt{2}}=\frac{1}{\sqrt{2}}\times\frac{\sqrt{2}}{\sqrt{2}}=\frac{\sqrt{2}}{2}.$$

$\sqrt{2}=1.414\ldots$ Without a calculator, $\dfrac{\sqrt{2}}{2}=\dfrac{1.414}{2}$ is much easier to find than $\dfrac{1}{\sqrt{2}}=\dfrac{1}{1.414}$

> Multiply numerator and denominator by the surd in the denominator

=== *Exercise 7* ===

1A Rationalise the denominators in these fractions, and simplify where possible:

    a $\dfrac{1}{\sqrt{6}}$ **b** $\dfrac{3}{\sqrt{7}}$ **c** $\dfrac{2}{\sqrt{6}}$ **d** $\dfrac{3}{\sqrt{3}}$ **e** $\dfrac{5}{\sqrt{10}}$

    f $\dfrac{20}{\sqrt{5}}$ **g** $\dfrac{1}{\sqrt{11}}$ **h** $\dfrac{1}{2\sqrt{2}}$ **i** $\dfrac{6}{\sqrt{12}}$ **j** $\dfrac{4}{\sqrt{20}}$.

2A Simplify:

    a $\dfrac{3}{2\sqrt{2}-\sqrt{2}}$ **b** $\dfrac{6}{2\sqrt{3}-\sqrt{3}}$ **c** $\dfrac{\sqrt{20}}{2\sqrt{5}-\sqrt{5}}$.

3A Express sin 45° and cos 45° as surds with rational denominators.

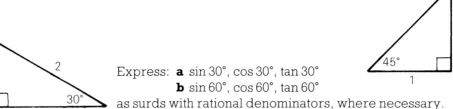

4A Express: **a** sin 30°, cos 30°, tan 30°

b sin 60°, cos 60°, tan 60°

as surds with rational denominators, where necessary.

5B a Using the quadratic formula $x = \dfrac{-b \pm \sqrt{(b^2 - 4ac)}}{2a}$, write down the roots of these

equations as surds in their simplest form:

(i) $x^2 - 4x + 1 = 0$ (ii) $x^2 + 2x - 17 = 0$ (iii) $2x^2 - 6x + 1 = 0$.

b Find the sum and product of the roots of each equation.

Conjugate surds

Using $(a - b)(a + b) = a^2 - b^2$, $(\sqrt{3} - 1)(\sqrt{3} + 1) = (\sqrt{3})^2 - 1^2 = 3 - 1 = 2$, a rational number.

$\sqrt{3} - 1$ and $\sqrt{3} + 1$ are *conjugate surds*—only a change of sign is involved.

Conjugate surds can be used to rationalise a denominator. For example,

$$\frac{1}{2 + \sqrt{3}} = \frac{1}{2 + \sqrt{3}} \times \frac{2 - \sqrt{3}}{2 - \sqrt{3}} = \frac{2 - \sqrt{3}}{2^2 - (\sqrt{3})^2} = \frac{2 - \sqrt{3}}{4 - 3} = 2 - \sqrt{3}.$$

6B Use conjugate surds to rationalise each denominator.

a $\dfrac{1}{\sqrt{5} - 1}$ **b** $\dfrac{1}{2 - \sqrt{3}}$ **c** $\dfrac{1}{\sqrt{2} + 1}$ **d** $\dfrac{2}{\sqrt{5} + \sqrt{3}}$ **e** $\dfrac{8}{\sqrt{6} - \sqrt{2}}$.

FRACTIONS

Reminders—

When you simplify $\dfrac{\overset{3}{\cancel{6}}}{\underset{4}{\cancel{8}}}$, you use the common factor of 6 and 8.

The same is true in algebra, $\dfrac{3x - 9}{x - 3} = \dfrac{3\overset{1}{\cancel{(x - 3)}}}{\underset{1}{\cancel{x - 3}}} = 3$.

═══════════════ *Exercise 8* ═══════════════

Copy and complete **1A**. Then simplify the expressions in the questions that follow. The golden rule is: **factorise first**, where possible.

1A a $\dfrac{\overset{1}{\cancel{8}}x}{\underset{2}{\cancel{16}}y} = \ldots$ **b** $\dfrac{4t - 8}{12} = \dfrac{\overset{1}{\cancel{4}}(t - 2)}{\underset{3}{\cancel{12}}} = \ldots$.

2A a $\dfrac{4x}{8y}$ **b** $\dfrac{10xy}{5x}$ **c** $\dfrac{a^2}{2ab}$ **d** $\dfrac{3cd}{6d}$ **e** $\dfrac{3k}{3k^2}$

3A a $\dfrac{2m + 6}{4}$ **b** $\dfrac{3n - 9}{6}$ **c** $\dfrac{2p + 4}{2}$ **d** $\dfrac{8t - 4}{4}$ **e** $\dfrac{x^2 + xy}{2x}$.

4A a $\dfrac{2}{2x+4}$ **b** $\dfrac{4}{4t-8}$ **c** $\dfrac{6}{3x+9}$ **d** $\dfrac{x^2}{x-xy}$ **e** $\dfrac{2a^2}{4a-6a^2}$.

Remember! Common factor, difference of squares, quadratic expression, cancellation.

5A a $\dfrac{2a+8}{a+4}$ **b** $\dfrac{3b-12}{2b-8}$ **c** $\dfrac{ax+bx}{a+b}$ **d** $\dfrac{x^2-y^2}{x-y}$ **e** $\dfrac{m^2-9}{m+3}$.

6A a $\dfrac{x^2+4x+3}{x+1}$ **b** $\dfrac{x^2+2x+1}{2x+2}$ **c** $\dfrac{y^2-5y+6}{3y-9}$ **d** $\dfrac{2z^2+12z+10}{2z+2}$.

7B a $\dfrac{u+3}{3u^2+3u-18}$ **b** $\dfrac{v^2-1}{v^2-5v+4}$ **c** $\dfrac{2p^2-2q^2}{p^2-2pq+q^2}$ **d** $\dfrac{y^4-1}{(y+1)(y^2+1)}$.

Multiplication and division

Exercise 9

Copy and complete **1A**. Then simplify the following questions:

1A a $\dfrac{2\cancel{x}^1}{\cancel{3}_1}\times\dfrac{\cancel{6}^2}{\cancel{x}_x}=\ldots$ **b** $\dfrac{5}{y}\div\dfrac{15}{y^3}=\dfrac{\cancel{15}^{\cancel{1}}}{\cancel{x}_x}\times\dfrac{\cancel{y}^{3\,y^2}}{\cancel{15}_3}=\ldots$.

2A a $\dfrac{x}{2}\times\dfrac{x}{3}$ **b** $\dfrac{y^2}{4}\times\dfrac{y}{2}$ **c** $\dfrac{x}{2}\times\dfrac{2}{3}$ **d** $\dfrac{y^2}{4}\times\dfrac{1}{y}$ **e** $\dfrac{z}{3}\times\dfrac{6}{z}$.

3A a $\dfrac{a}{3}\div\dfrac{a}{4}$ **b** $\dfrac{b^2}{2}\div\dfrac{b}{5}$ **c** $\dfrac{c^3}{4}\div\dfrac{c}{2}$ **d** $\dfrac{2d}{3}\div\dfrac{d}{4}$ **e** $\dfrac{x^4}{4}\div\dfrac{x^2}{2}$.

4B Calculate the area of each shape in terms of x:

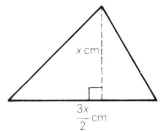

a Rectangle — $\dfrac{x}{3}$ cm, $\dfrac{x}{2}$ cm

b x cm, $\dfrac{3x}{2}$ cm

c Kite — Diagonal lengths $\dfrac{2x}{3}$ cm and $\dfrac{3x}{2}$ cm

5B The area of a rectangle is $x^2+7x+12$ cm². Its length is $x+4$ cm. Find its width.

Addition and subtraction

Exercise 10

Copy and complete **1A**. Then do the other questions in the same way.

1A a $\dfrac{3}{4}+\dfrac{5}{6}$ (lcm of 4 and 6 is 12)

$=\dfrac{3\times3+2\times5}{12}$

$=\ldots$

b $\dfrac{3x}{5}-\dfrac{2x}{7}$

$=\dfrac{7\times3x-5\times2x}{35}$ (lcm of 5 and 7 is 35)

$=\ldots.$

2A a $\dfrac{2}{3}+\dfrac{3}{4}$ **b** $\dfrac{1}{6}+\dfrac{1}{2}$ **c** $\dfrac{4}{5}-\dfrac{3}{10}$ **d** $\dfrac{1}{2}-\dfrac{2}{5}$ **e** $\dfrac{7}{8}+\dfrac{2}{3}$.

3A a $\dfrac{x}{3}+\dfrac{x}{4}$ **b** $\dfrac{y}{2}-\dfrac{y}{3}$ **c** $\dfrac{2u}{3}-\dfrac{u}{5}$ **d** $\dfrac{3v}{4}+\dfrac{2v}{3}$ **e** $\dfrac{w}{10}-\dfrac{2w}{5}$.

4A a $\dfrac{3}{x^2}-\dfrac{1}{x^2}$ **b** $\dfrac{2}{y^2}+\dfrac{3}{y}$ **c** $\dfrac{4}{x}-\dfrac{5}{y}$ **d** $\dfrac{x}{y}+\dfrac{y}{x}$ **e** $\dfrac{xy}{z}-\dfrac{1}{z^2}$.

5A Copy and complete **a**, then go on to the other parts:

a $\dfrac{2n-1}{2}-\dfrac{n-2}{3}$ **b** $\dfrac{x+1}{2}+\dfrac{x-1}{3}$ **c** $\dfrac{a+3}{4}+\dfrac{a-2}{6}$

$=\dfrac{3(2n-1)-2(n-2)}{6}$ **d** $\dfrac{2v+3}{4}+\dfrac{v+1}{2}$ **e** $\dfrac{w-2}{5}+\dfrac{2w+3}{2}$

$=\dfrac{6n-3-2n+4}{6}$ (Why?)   **f** $\dfrac{x+3}{4}-\dfrac{x+1}{3}$ **g** $\dfrac{y-1}{2}-\dfrac{y-3}{3}$.

$=\ldots$

6B Copy and complete **a**, then go on to the other parts:

a $\dfrac{2}{x+2}-\dfrac{1}{x+3}$ **b** $\dfrac{4}{x+1}+\dfrac{1}{x+2}$ **c** $\dfrac{5}{x-2}+\dfrac{3}{x+2}$

$=\dfrac{2(x+3)-1(x+2)}{(x+2)(x+3)}$ **d** $\dfrac{3}{a+3}-\dfrac{2}{a+2}$ **e** $\dfrac{1}{x}-\dfrac{1}{x-1}$

$=\dfrac{2x+6-x-2}{(x+2)(x+3)}$ **f** $\dfrac{1}{x^2}-\dfrac{1}{x^2+1}$ **g** $\dfrac{x}{x+5}-\dfrac{x}{x-5}$.

$=\ldots$

Equations with fractions

Reminder—Multiply each side by the lcm of the denominators.

═══════════════ *Exercise 11* ═══════════════

1A Copy and complete **a**, **b** and **c**. Then solve the other equations.

a $\dfrac{x}{2}=-3$ **b** $\dfrac{6}{x-1}=2$ **c** $\dfrac{x-1}{3}-\dfrac{x+1}{4}=5$

lcm is 2   lcm is $x-1$ lcm is 12

$2\times\dfrac{x}{2}=2\times(-3)$ $x-1\times\dfrac{6}{x-1}=2(x-1)$ $12\left(\dfrac{x-1}{3}-\dfrac{x+1}{4}\right)=12\times5$

$x=\ldots$ $6=\ldots$ $4(x-1)-3(x+1)=60$.

                                                                   $\cdots\qquad\cdots$

2A a $\dfrac{x}{3}=5$ **b** $\dfrac{y}{2}=-1$ **c** $\dfrac{m+2}{4}=3$ **d** $\dfrac{n-4}{3}=\dfrac{1}{2}$.

3A a $\dfrac{6}{x} = 2$ **b** $\dfrac{3}{y+1} = 1$ **c** $\dfrac{2}{2z-1} = -3$ **d** $\dfrac{1}{2x-4} = \dfrac{1}{3}$.

4A a $\dfrac{x}{4} + \dfrac{x}{5} = 1$ **b** $\dfrac{2x}{3} - \dfrac{3x}{4} = 0$ **c** $\dfrac{y+2}{5} + \dfrac{y}{3} = 2$ **d** $\dfrac{y}{2} - \dfrac{y-1}{4} = -1$.

5B a $\dfrac{x+3}{6} - \dfrac{x+1}{2} = 4$ **b** $\dfrac{2x-1}{3} - \dfrac{2x-1}{5} = 2$ **c** $\dfrac{y+1}{2} - \dfrac{y-2}{3} = \dfrac{1}{4}$.

6B a $\dfrac{x}{2} + \dfrac{2}{x} + 2 = 0$ **b** $\dfrac{1}{x-1} - \dfrac{2}{x+1} = 0$ **c** $\dfrac{1}{x+2} - \dfrac{1}{x} = 2$.

Here's how you can tell them in a split second what the sum of $\dfrac{1}{1\times2} + \dfrac{1}{2\times3} + \dfrac{1}{3\times4} + \ldots$ is to any number of terms.

First, practise finding the sum of the first 3 terms using the reciprocal $(1/x)$ key on your calculator for $\frac{1}{2}$, $\frac{1}{6}$ and $\frac{1}{12}$, like this:

| 2 | 2ndF | 1/x | + | 6 | 2ndF | 1/x | = | + | 12 | 2ndF | 1/x | = |

How long did it take? Without any working, the answer is $\frac{3}{4}$. Why?

Here is the secret: $S = \dfrac{1}{1\times2} + \dfrac{1}{2\times3} + \dfrac{1}{3\times4}$

$$= \left(1 - \frac{1}{2}\right) + \left(\frac{1}{2} - \frac{1}{3}\right) + \left(\frac{1}{3} - \frac{1}{4}\right)$$

$$= 1 - \frac{1}{4} \quad \text{Why?}$$

$$= \frac{3}{4}.$$

1 *Write down* the sum of $\dfrac{1}{1\times2} + \dfrac{1}{2\times3} + \ldots + \dfrac{1}{5\times6}$. Then write it out as above, to check.

2 Can you write down the sum of $\dfrac{1}{1\times2} + \dfrac{1}{2\times3} + \ldots + \dfrac{1}{99\times100}$?

3 a Simplify $\dfrac{1}{n} - \dfrac{1}{n+1}$ **b** Write down the sum of $\dfrac{1}{1\times2} + \dfrac{1}{2\times3} + \ldots + \dfrac{1}{n(n+1)}$.

4 Find the sum of $\dfrac{1}{1\times3} + \dfrac{1}{2\times4} + \dfrac{1}{3\times5} + \dfrac{1}{4\times6} + \ldots + \dfrac{1}{48\times50}$.

CHECK-UP ON **INDICES AND FRACTIONS**

1 Simplify:

a $x^7 \times x^3$ **b** $x^{12} \div x^{10}$ **c** $x^{1/2} \times 2x^{3/2}$ **d** $x^{4/3} \div x^{1/3}$ **e** $(x^{1/2})^{-2/3}$.

2 Given $x = 16$ and $y = 9$, find the value of:

a $x^{3/4} + y^{1/2}$ **b** $x^{3/2} - y^{3/2}$ **c** $x^{-1/2} + y^{-1/2}$ **d** $(x+y)^{1/2}$ **e** $(x-y)^0$.

3 Express as a sum of powers of n; for example, $\dfrac{1 - 2n^3}{n^2} = \dfrac{1}{n^2} - 2n$.

a $\dfrac{n^2 - 1}{n}$ **b** $\dfrac{n^2 + 2}{n}$ **c** $\dfrac{1 + \sqrt{n}}{n}$.

4 Write as single fractions; for example, $1 + \dfrac{1}{x} = \dfrac{x+1}{x}$.

a $x^{-1} + x^{-2}$ **b** $x + 2x^{-1}$ **c** $3x^{-2} - x$.

5 Draw the graphs of $y = 5^x$ and $y = 5^{-x}$, from $x = -1$ to 2.

6 Find x, given: **a** $5^x = 25^2$ **b** $9 \times 3^x = 3^6$ **c** $2^x = 4^3 \times 8^2$.

7 Simplify:

a $\sqrt{8}$ **b** $\sqrt{400}$ **c** $3\sqrt{12}$ **d** $\sqrt{80}$ **e** $\sqrt{3} \times \sqrt{18}$ **f** $\sqrt{\dfrac{16}{121}}$.

8 Write in a simpler form:

a $\sqrt{3} + 4\sqrt{3}$ **b** $5\sqrt{7} - 2\sqrt{7}$ **c** $\sqrt{2}(\sqrt{2} - \sqrt{8})$ **d** $\sqrt[3]{54}$.

9 Rationalise the denominators:

a $\dfrac{1}{\sqrt{5}}$ **b** $\dfrac{2}{\sqrt{6}}$ **c** $\dfrac{4}{3\sqrt{2}}$ **d** $\dfrac{6}{\sqrt{5} - \sqrt{2}}$ **e** $\dfrac{4}{1 - \sqrt{3}}$.

10 Simplify:

a $\dfrac{2x}{14}$ **b** $\dfrac{3y - 12}{3}$ **c** $\dfrac{4z - 2}{6z - 3}$ **d** $\dfrac{a + 2}{a^2 - 4}$ **e** $\dfrac{(b + 1)^2}{b^2 + 2b + 1}$.

11 Express as single fractions:

a $\dfrac{x^2}{3} \times \dfrac{3}{x}$ **b** $6t \div \tfrac{1}{2}t$ **c** $\dfrac{7t}{5} - \dfrac{2t}{3}$ **d** $\dfrac{p - 1}{2} - \dfrac{p + 1}{3}$.

12 a Express $\dfrac{1}{n + 2} - \dfrac{1}{n + 3}$ as a single fraction.

b Calculate the sum of 10 terms of $\dfrac{1}{2 \times 3} + \dfrac{1}{3 \times 4} + \dfrac{1}{4 \times 5} + \ldots$.

13 Solve:

a $\dfrac{2x}{3} = 4$ **b** $\dfrac{6}{x - 1} = 2$ **c** $\dfrac{x + 3}{3} + \dfrac{x}{4} = 1$ **d** $\dfrac{x + 3}{4} - \dfrac{x - 1}{2} = 0$.

THE GRADIENT OF A STRAIGHT LINE

How can we measure the gradient of the flight–path of this climbing aircraft?

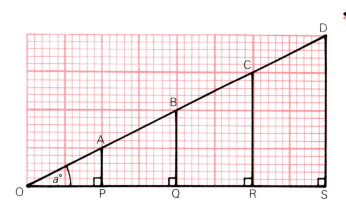

From similar triangles, $\dfrac{PA}{OP} = \dfrac{QB}{OQ} = \dfrac{RC}{OR} = \dfrac{SD}{OS} = \dfrac{200\,m}{400\,m} = \dfrac{1}{2}$.

For *every* point on the flight–path, $\dfrac{\text{vertical distance above O}}{\text{horizontal distance from O}} = \dfrac{1}{2}$.

The gradient of the flight–path $= \frac{1}{2}$.

> The *gradient* of a slope $= \dfrac{\text{vertical distance between each end}}{\text{horizontal distance between each end}}$.

Exercise 1

1A Calculate the gradient of each ladder:

a

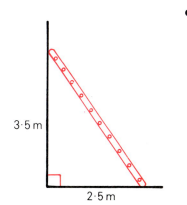

16 ft

8 ft

b

3·5 m

2·5 m

c

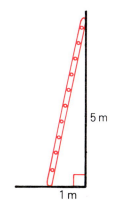

5 m

1 m

177

2A Calculate the gradient of each staircase:

a

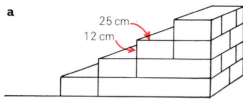

25 cm
12 cm

b

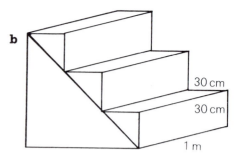

30 cm
30 cm
1 m

3A

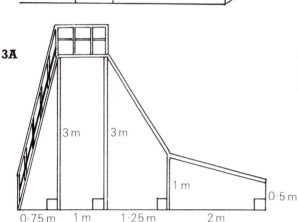

3 m   3 m

1 m

0·5 m

0·75 m   1 m   1·25 m   2 m

Calculate the gradient of each of the four parts of the slide. Make a sketch of the slide first.

4A a A road rises 5 metres over a horizontal distance of 125 m.
b A railway line rises 1 metre over a horizontal distance of 250 m.
Calculate their gradients in decimal form.

5B Calculate the gradient between pairs of points A, B and C, D on these map contour lines:

a

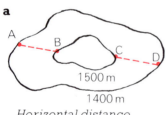

A
B
C
D
1500 m

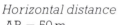

1400 m

Horizontal distance
AB = 50 m
CD = 75 m

b

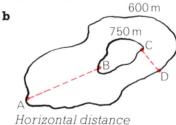

600 m
750 m
C
B
D
A

Horizontal distance
AB = 600 m
CD = 150 m

c

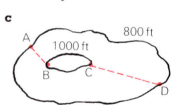

800 ft
A
1000 ft
B
C
D

Horizontal distance
AB = 120 feet
CD = 300 feet.

6B Calculate the gradient of the seesaw when:
a one end is on the ground (careful!)
b it is horizontal.

In mathematics, the *gradient* of AB

$$= \frac{\text{change in } y \text{ from A to B}}{\text{change in } x \text{ from A to B}}$$

$$= \frac{y\text{-step}}{x\text{-step}}$$

$$= \frac{MB}{AM} = \frac{3}{6}, \text{ or } \frac{1}{2}.$$

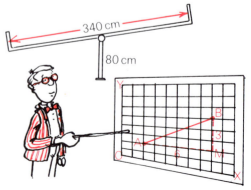

340 cm
80 cm

Note The letter m is often used for the gradient.
For the gradient of AB $= \frac{1}{2}$, we write $m_{AB} = \frac{1}{2}$.

Example 1

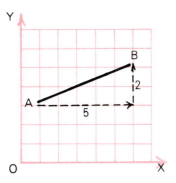

Gradient of AB $= \frac{2}{5}$.

AB slopes *up* from left to right.
It has a *positive gradient*.

Example 2

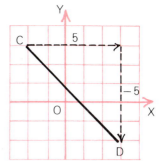

Gradient of CD $= \frac{-5}{5} = -1$.

CD slopes *down* from left to right.
It has a *negative gradient*.

Example 3

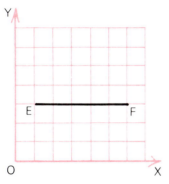

$m_{EF} = \frac{0}{5} = 0$.

EF is parallel to the *x*–axis.

Example 4

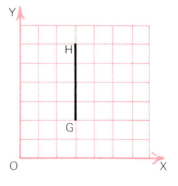

$m_{GH} = \frac{5}{0}$, which has no value.

GH is parallel to the *y*–axis.

Exercise 2

1A Calculate the gradient of each line. For example, $m_{AB} = \frac{3}{3} = \ldots$.

a

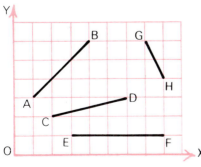

b

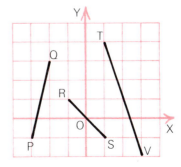

2A a Plot these pairs of points, and calculate the gradient of the line joining each pair:

(i) A(3, 1) and B(6, 2)   (ii) O(0, 0) and C(2, 4)   (iii) D(-2, -2) and E(2, 2)
(iv) F(-2, 0) and G(1, 1)   (v) H(-3, 3) and K(-4, -2)   (vi) M(3, -4) and N(6, -4).

b From the gradients of AB and FG, what can you say about these two lines?

3A a Calculate the gradients of the sides of the quadrilaterals.

b What type of quadrilateral is:
(i) KLMN   (ii) EFGH?

c What do you notice about the gradients of the sides of the kite?

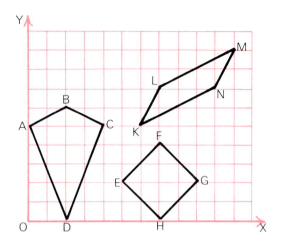

4A For which line is the gradient: **a** positive   **b** negative   **c** zero   **d** none of these?

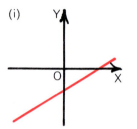 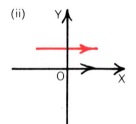 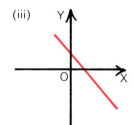 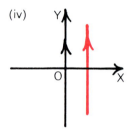

5B On squared paper, draw lines through O with gradients:
a 1   **b** 3   **c** −1   **d** $\frac{1}{2}$.   Which line has the steepest slope?

6B On squared paper, draw lines with gradient 1, through:
a $(0, 2)$ **b** $(0, 0)$ **c** $(0, -2)$.

7B A is the point $(-1, -1)$, B$(1, 2)$ and C$(5, 8)$.
a Calculate: (i) m_{AB} (ii) m_{BC} (iii) m_{AC}.
b What can you say about the three points?

P is the point (a, b). OP is rotated through 90° to OQ.
(i) Write down the coordinates of Q in terms of a and b.
(ii) Show that $m_{OP} \times m_{OQ} = -1$.
(iii) Write down the gradients of lines perpendicular to lines with gradients $2, \frac{1}{3}, \frac{5}{8}$ and -4.

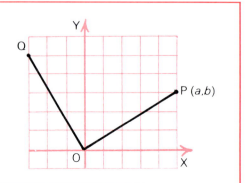

$m_{AB} = \dfrac{MB}{AM} = \tan \theta°$, where $\theta°$ is the angle
AB makes with OX.
Use your calculator to calculate θ.
Calculate the sizes of the angles made by the
lines in question **1** of Exercise **2** with OX.

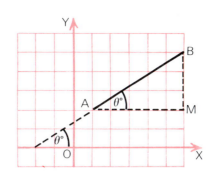

THE GRADIENT OF A CURVE

Finding the gradient of a straight line
is easy. But a curve is always changing
direction. How can we measure its gradient?

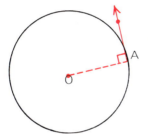

When the hammer–thrower lets go, the
hammer flies off *at a tangent* to the circle.

The gradient of a curve at A is the gradient of the tangent at A.

Examples   (i)

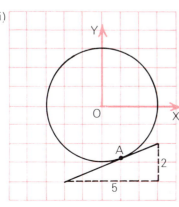

(ii)

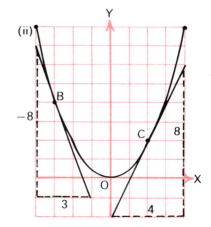

The gradient of the tangent to the circle
at $A = \frac{2}{5}$.

In the same way, the gradient of the tangent
to the parabola:
(i) at $B = -\frac{8}{3}$ (ii) at $C = \frac{8}{4} = 2$.

GRADIENTS, GRAPHS AND NO–GO REGIONS

1A Calculate the gradients of the tangents to the circles at the points marked:

a

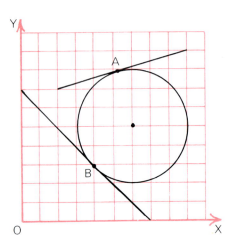

b

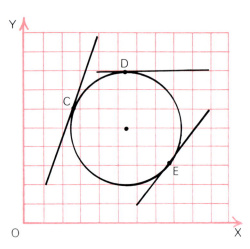

2A Trace this curve, and draw a tangent at each numbered point. Place your tracing on squared paper. By reading off distances in the *x*- and *y*-directions, calculate the gradients of the tangents at the numbered points.

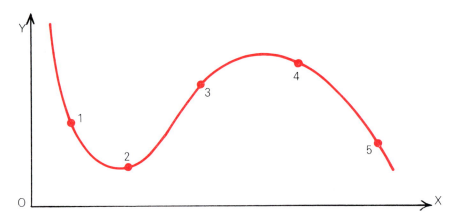

3A a Copy and complete:

x	-4	-3	-2	-1	0	1	2	3	4
$\frac{1}{2}x^2$	8	$4\frac{1}{2}$	2	$\frac{1}{2}$	0				

b Draw the graph of $y = \frac{1}{2}x^2$ from $x = -4$ to 4, using the scales shown.

c Draw the tangents to the parabola at $x = -4$ and $x = 2$.

d Estimate the gradients of the tangents at these points.

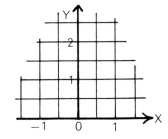

4B a Make a table of values for the hyperbola $y = \dfrac{12}{x}$, for $x = 1, 1 \cdot 5, 2, 3, 4, 6, 8, 10, 12$.

b Draw the graph of the hyperbola, taking 1 unit to 1 cm on each axis.

c Draw the tangents to the hyperbola at the points where $x = 2$, $x = 4$ and $x = 8$, and find their gradients.

5B a Copy and complete this table of values:

x	-2	$-1 \cdot 5$	-1	$-0 \cdot 5$	0	$0 \cdot 5$	1	$1 \cdot 5$	2
4	4	4	4	4		4			
x^2	4	$2 \cdot 25$	1			$0 \cdot 25$			
$4 - x^2$	0	$1 \cdot 75$	3			$3 \cdot 75$			

b Draw the graph of the parabola $y = 4 - x^2$ from $x = -2$ to 2.
Use 2 mm squared paper, and 1 unit to 2 cm on each axis.

c From your graph, find the gradients of the tangents at the points where $x = -1 \cdot 5$, $x = -0 \cdot 5$ and $x = 1$.

A curve with a secret
Use the exponential key $\boxed{e^x}$ on your calculator to help you to draw the graph of $y = e^x$, for $x = 0, 0 \cdot 5, 1 \cdot 5, \ldots, 4$. Investigate the gradients of the tangents at several points on the graph. What can you discover?

GRADIENT AS A MEASURE OF RATE OF CHANGE

The old banger is having its speed checked. Distance is then graphed against time.

(i) The *average speed* is the rate of change of distance with time

$$= \frac{\text{distance}}{\text{time}}$$

$$= \frac{100 \, \text{m}}{5 \, \text{s}}$$

$$= 20 \, \text{m/s}.$$

((i) The gradient of the line $= \dfrac{20}{1} = 20$. (See diagram.)

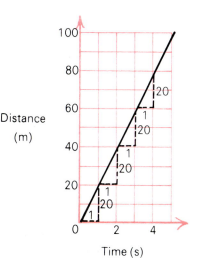

The average speed corresponds to the gradient of the line.

1A a Using the scales shown above, draw the distance—time graph for:

Time (seconds)	0	1	2	3	4
Distance (metres)	10	40	70	100	130

b Calculate the average speed in m/s.

c Read off the gradient of the line in your graph. Compare your answers to **b** and **c**.

2A a Calculate the gradient of AB, and hence write down the average speed from A to B.

b Repeat **a** for AC.

c *The actual speed at A corresponds to the gradient of the tangent at A.*
Estimate this speed.

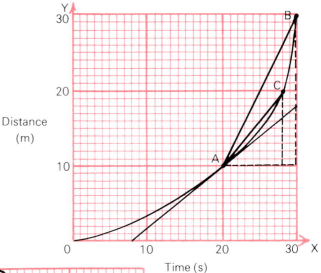

Distance (m)

Time (s)

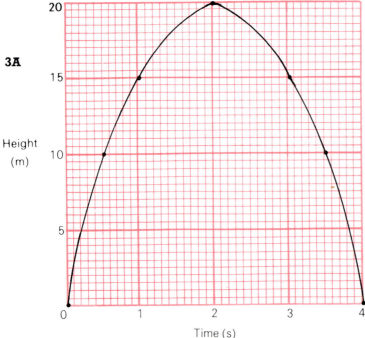

3A

Height (m)

Time (s)

A ball is thrown straight up in the air. Its height is plotted against time. Use the gradient of a tangent to estimate the speed of the ball:

a after (i) 1 s (ii) 2 s (iii) 2·5 s

b at the start.

4A The graph shows the typical growth pattern of a boy from 0 to 16.

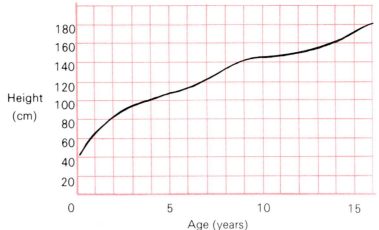

Height (cm)

Age (years)

a Estimate his rate of growth, that is the rate of change of his height, in cm per year, at age:
(i) 2   (ii) 12   (iii) 15.
(Use the gradient of a tangent.)

b Calculate his *average* annual rate of growth over the 16 years.

5B A new car's petrol consumption is being checked.

Petrol used (gallons)	0	5	10	15	20
Miles travelled from start	0	120	260	350	500

a Draw a graph.
b Calculate its petrol consumption in miles per gallon for:
(i) each stage of his journey   (ii) the whole journey.

Reaching the limit

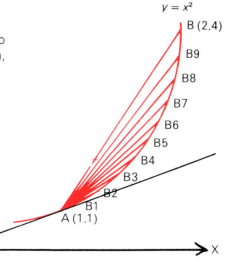

Calculate the gradient of the tangent to the parabola $y = x^2$ at the point A(1, 1), as follows.
Copy and complete:

B(2, 4); $m_{AB} = \dfrac{4-1}{2-1} = 3$

B9(1·9, 1·9²); $m_{AB9} = \dfrac{3·61-1}{1·9-1} = 2·9$

B8(1·8, 1·8²); $m_{AB8} = \ldots$

. . . .

B1(1·1, 1·1²); $m_{AB1} = \ldots .$

As $B \rightarrow A$, gradient of chord AB$n \rightarrow$ gradient of tangent at A.
So gradient of tangent at A = ...

Challenge   A is (1, 1) and B is $(1+h, (1+h)^2)$.
Find an expression for the gradient of AB in terms of h, in its simplest form.
Hence find the gradient of the tangent at A.

THE EQUATION OF A STRAIGHT LINE, $y = mx + c$

GRADIENTS, GRAPHS AND NO-GO REGIONS

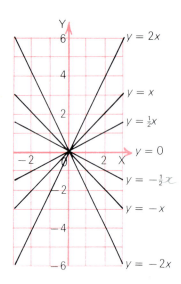

In Chapters **1** and **4** you discovered that lines through the origin had equations like $y = x$, $y = 3x$, $y = -2x$, $y = -\frac{1}{3}x$, . . . , $y = ax$. Check that for each line in the diagram, a is just its gradient.

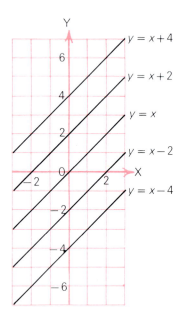

You also discovered that all straight lines (not parallel to the y-axis) had equations of the form $y = ax + b$. Check that for each line in the diagram, b tells you how far from the origin the line cuts the y-axis.

It is traditional to write this equation as $y = mx + c$, where m = gradient, and c = y-intercept.

Example 1   Find the equations of these two lines.

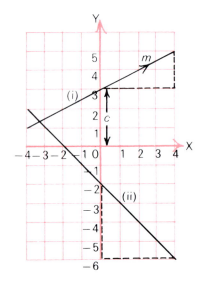

Line (i)

$m = \frac{2}{4} = \frac{1}{2}$, and $c = 3$.
Using $y = mx + c$, the equation of the line is $y = \frac{1}{2}x + 3$, or $2y = x + 6$.

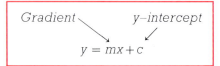

Gradient   *y–intercept*

$$y = mx + c$$

Line (ii)

$m = \frac{-4}{4} = -1$, and $c = -2$.

The equation of the line is $y = -x - 2$.

Example 2 Find the equation of the line through A(0·5, 8) and B(2, 14).

$$m_{AB} = \frac{MB}{AM} = \frac{14-8}{2-0·5} = \frac{6}{1·5} = 4.$$

Using $y = mx + c$, the equation of AB is
$y = 4x + c$.
(2, 14) lies on AB, so $14 = 8 + c$
$$c = 6.$$

The equation of AB is $y = 4x + 6$.

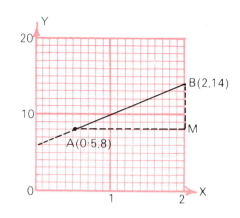

Exercise 5

1A Write down the gradients and y–intercepts (distances from the origin to the points where the lines cut the y–axis) of the lines with equations:

a $y = 2x + 5$ **b** $y = \frac{1}{2}x - 4$ **c** $y = -x + 1$ **d** $y = -8x$.

2A Write down the gradient of each line, and the point where it cuts the y–axis:

a $y = x + 2$ **b** $y = -3x - 1$
c $y = 5x$ **d** $y = 4$.

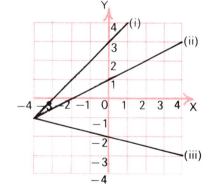

3A On squared paper, draw these pairs of lines, and write down their gradients and y–intercepts:

a $y = x$ and $y = x + 4$ **b** $y = 2x$ and $y = 2x - 5$
c $y = -x$ and $y = -x + 3$ **d** $y = -2x$ and $y = -2x - 1$.

4A Write down the gradient and y–intercept of each line, and hence find its equation.

5A Write down the equations of the lines through O with gradients:

a 5 **b** $\frac{1}{4}$ **c** $-\frac{1}{2}$ **d** 0 **e** $\frac{3}{2}$.

6A Write down the equations of the lines with gradient 4, passing through:

a $(0, 2)$ **b** $(0, -2)$ **c** $(0, \frac{1}{2})$ **d** $(0, 0)$.

7B Use the method of worked *Example 2* above to find the equations of the lines joining these pairs of points:

a A(3, 6) and B(5, 8)   **b** C(2, 4) and D(10, 6)   **c** E(2, 1) and F(5, 7)
d G(2, 1) and H(8, 3)   **e** P(0·5, 1·5) and Q(2, 6)   **f** R(0·1, 10) and S(0·9, 20).

8B a Rearrange each equation in the form
$y = mx + c$.

  b Write down the gradient and the y-intercept of the line it represents.

(i) $x + y = 6$ (ii) $-2x + y = 3$
(iii) $4y = 8x + 1$ (iv) $x + 2y = -4$
(v) $3x + 2y = 6$ (vi) $x + y + 1 = 0$.

9B In which of these pairs are the lines parallel?

a $\left.\begin{array}{l} y = 2x - 5 \\ y - 2x = 1 \end{array}\right\}$ **b** $\left.\begin{array}{l} x + y = 4 \\ x - y = 4 \end{array}\right\}$ **c** $\left.\begin{array}{l} 2x - 3y = 6 \\ 3x - 2y = 6 \end{array}\right\}$ **d** $\left.\begin{array}{l} 3x + 5y + 7 = 0 \\ 3x + 5y - 7 = 0 \end{array}\right\}$.

LINE OF BEST FIT

Experimenters at work

Mandy is investigating the effect of different pressures on the boiling point of water. She records her measurements, and plots the seven points corresponding to the pairs of measurements. They lie close to a straight line, apart from the last one—a bad measurement.

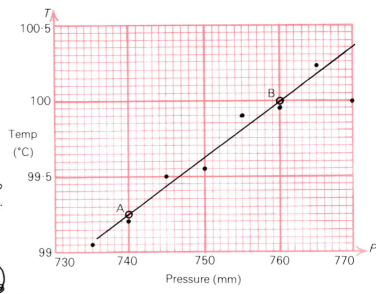

Pressure (P mm)	735	740	745	750	755	760	765	770
Boiling point ($T°$C)	99·05	99·20	99·50	99·55	99·90	99·95	100·25	100·00

It wasn't easy, but eventually Mandy managed to draw a straight line 'balanced between all the points'. 'Now for the equation of my *best-fitting* straight line', she thought.

(i) 'The equation is like $y = mx + c$, but with my letters, $T = mP + c$.'
(ii) 'I'll choose two points on my line, spaced well apart: A(740, 99·25) and B(760, 100). Now I've a choice of two ways of finding the equation of AB.'

Method 1

Using the gradient of the graph

$$m_{AB} = \frac{100 - 99 \cdot 25}{760 - 740}$$

$$= 0 \cdot 04 \text{ (2 dec. places)}$$

$$T = mP + c$$

So $T = 0 \cdot 04P + c$

(760, 100) lies on the line,

so $100 = 0 \cdot 04 \times 760 + c$

$$c = 100 - 30 \cdot 4$$

$$= 69 \cdot 6$$

The equation of the line is

$$T = 0 \cdot 04P + 69 \cdot 6.$$

Method 2

Using pairs of equations

$$T = mP + c$$

When $P = 760$, $T = 100$, so $100 = 760m + c$

When $P = 740$, $T = 99 \cdot 25$, so $99 \cdot 25 = 740m + c$

Subtract $\quad 0 \cdot 75 = 20m$

$$m = \frac{0 \cdot 75}{20}$$

$$\doteqdot 0 \cdot 04.$$

Put $m = 0 \cdot 04$ in the first equation.

$$100 = 760 \times 0 \cdot 04 + c$$

$$c = 100 - 30 \cdot 4$$

$$c = 69 \cdot 6.$$

The equation of the line is

$$T = 0 \cdot 04P + 69 \cdot 6.$$

'Either way', thought Mandy, 'it's hard work'.

Exercise 6

1A To check the maximum speed of a new plane a test–pilot fills in this table:

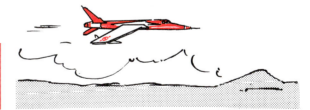

Time (T seconds)	1	2	3	4
Distance (D metres)	410	540	710	855

a Plot the points (1, 410), (2, 540), . . . , using the scales shown.

b Draw what you consider to be the best–fitting straight line.

c Choose two well–separated points on your line to find the equation of the line in the form $D = mT + c$.

d Calculate the plane's average speed.

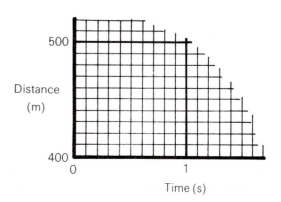

Distance (m)

500

400

0   1

Time (s)

2A Vic checks the rate of fall of the water level in the tank.

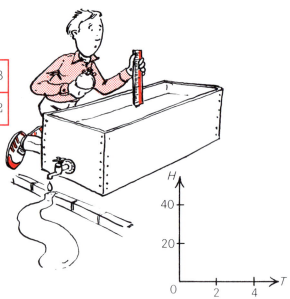

Time (T min)	3	5	6	10	13
Water depth (H cm)	86	72	68	52	32

a Plot the points on 2 mm squared paper. (See axes.)
b Draw your best-fitting straight line.
c Calculate the gradient of the line. Write down the rate of fall of the water level (cm/min).
d Find the equation of the line. Estimate the time needed to empty the tank.

3A Peter is given some pieces of iron, and asked to find the density of iron (the mass of 1 cm³). So he measures the volume and mass of each piece.

Volume (V cm³)	5	11	15	23	36	49
Mass (M g)	45	90	120	175	290	390

a Plot the points on 2 mm squared paper, taking the V–axis horizontal (scale 10 cm³ to 2 cm) and the M–axis vertical (scale 100 g to 2 cm).
b Draw your best–fitting straight line—why must it pass through (0, 0)?
c Choose two well separated points on the line, and calculate its gradient. Write down the density of iron.

4A Rory was in real trouble. Breathalysed. The table shows his blood alcohol level over the next five hours.

Time (T hours)	0	1	2	3	4	5
Alcohol (A mg/ml)	194	180	165	152	139	125

a Plot the points on 2 mm squared paper, and draw your best-fitting straight line.
b Calculate the average rate of fall of his alcohol level in mg/ml per hour.
c Find the equation of the line, and estimate how long it will take for the alcohol level to fall to 50 mg/ml.

5B An experiment to find a toy car's acceleration down a slope.

The car is released at A, and passes through electronic gates at B and C. C can be moved to vary the distance BC. A computer wired to gates B and C gives a 'read out' of the car's speed at C and the time from B to C.

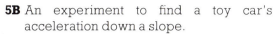

Time (T s)	1·1	2·1	3	3·9	5·2	5·9
Speed (S cm/s)	5	12	15	22	26	30

a Plot the points, and draw a line of best fit.
b The gradient of the line gives the rate of change of speed with respect to time, that is the *acceleration* of the car. Estimate it.

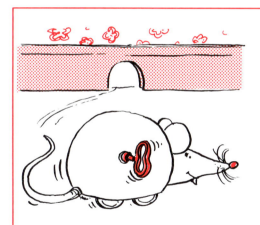

The length of time a clockwork toy runs depends on the number of turns the key is given. Is the relation linear? Set up an experiment to find out.

Meena takes 80 seconds to reach the ground floor in a lift. After 32 seconds she passes the 21st floor. Assuming that the lift descends steadily, without stops, find at which floor she entered the lift. *HINT* Draw a graph.

NO–GO REGIONS OF THE PLANE

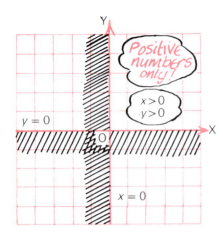

For all points (x, y) in the *unshaded region*, $x > 0$ and $y > 0$. The shaded regions are the 'no–go' areas.

The line $y = x$ divides the plane into two regions: (i) For all points, $y > x$. (ii) For all points, $y < x$.

Example   Leave these regions clear, by shading the no–go areas:

a $y > x + 2$

b $3x + 2y < 12$

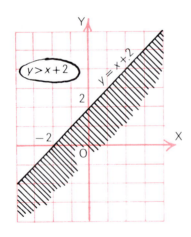

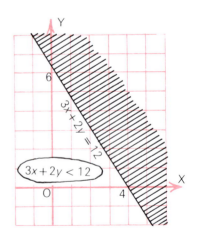

(i) Draw the line $y = x + 2$.

(ii) Decide where $y > x + 2$.
Test $(0, 0)$ in the inequation. $0 > 0 + 2$; not true, so $(0, 0)$ is in the shaded no–go region.

(i) Draw the line $3x + 2y = 12$.
(It cuts the axes at $(0, 6)$ and $(4, 0)$).

(ii) Decide where $3x + 2y < 12$.
Test $(0, 0)$ in the inequation. $0 + 0 < 12$; true, so $(0, 0)$ is in the clear region.

1A Use inequations to describe the clear regions:

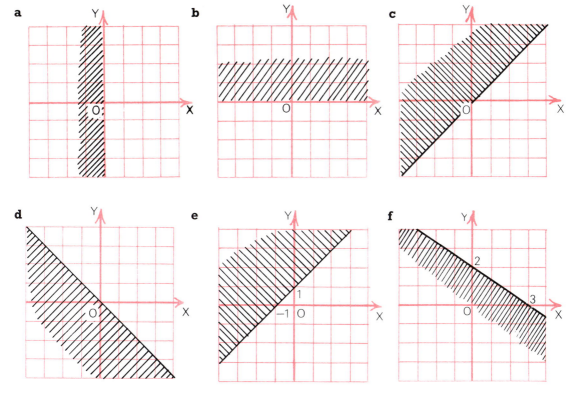

2A On squared paper leave unshaded the regions where for all points (x, y):

   a $y > x + 1$ **b** $y < -x + 4$ **c** $x + y > 5$ **d** $5x - 2y > 10$.

3A Write down a set of inequations which define each clear region:

          a   **b**   **c**

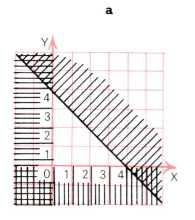
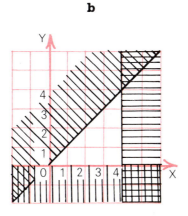
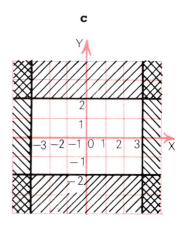

4A a On the same diagram, show clearly the regions where:
(i) $y < x+1$ (ii) $y > x-1$ (iii) $x < 2$ (iv) $x > -2$.
b What is the shape of the resulting unshaded region?

5A Repeat question **4A** for the inequations.
(i) $x > 0$ (ii) $y > 0$ (iii) $y < 3$ (iv) $y < 5-x$.

6B a Draw the three straight lines through:
(i) $(-1, -1)$ and $(2, 5)$ (ii) $(-1, -1)$ and $(2, -4)$ (iii) $(2, 0)$, parallel to the y−axis.
b Find three inequations which define the triangular region bounded by the lines.

MAXIMUM AND MINIMUM VALUES OF '$ax + by$'

The clear region, *including its boundaries,* is defined by this set of inequations:
$x \geqslant 0$, $y \geqslant 0$,
$x + 2y \leqslant 9$, $3x + y \leqslant 12$.

To find the maximum or minimum value of $x + y$ in the region, take a 'searchline' $x + y = k$.
Draw the line for $k = 0, 1, 2, 3, \ldots$
until you find its maximum value 6, when it passes through the vertex $(3, 3)$. Its minimum value is 0, at the origin.
In practice, you can slide your ruler along parallel to the line $x + y = 0$ to find the points you want.

===================== *Exercise 8* =====================

1A Place your ruler on the line $x + y = 3$. Slide it parallel to the line to find the maximum and minimum values of $x + y$ in the clear region (including the boundary lines.)

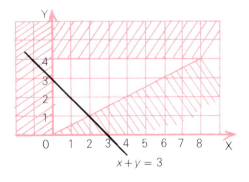

2A Repeat question **1A** for the regions defined by the set of inequations:

   a $x \geqslant 0$, $y \geqslant 0$, $x \leqslant 5$ and $y \leqslant 6$.

   b $x \geqslant 0$, $y \geqslant 0$ and $2x + 3y \leqslant 12$.

   c $x \geqslant 0$, $y \geqslant 0$, $2x + y \leqslant 6$ and $x + 2y \leqslant 6$.

3A a Show the region defined by: $x \geqslant 0$, $y \geqslant 0$, $x \leqslant 5$ and $y \leqslant 4$.

   b Draw the line $2x + y = 8$.

   c Use the searchline method to find the maximum and minimum values of $2x + y$ in the region.

4A The equations of five straight lines bounding a clear region are shown.

   Use the searchline $x + 3y = k$ to find the maximum and minimum values of $x + 3y$ in the region.

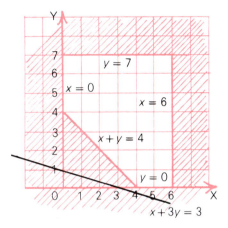

5B a Show the region defined by $x \geqslant 1$, $x \leqslant 4$, $y \geqslant 0$, $x + y \leqslant 6$ and $2x + y \geqslant 3$.

   b Show that the maximum value of $4x + y$ in the region is 18, and the minimum value is 5.

6B Repeat question **5B** for the maximum and minimum values of $2x + 4y$ in the same region.

CHECK-UP ON **GRADIENTS, GRAPHS AND NO–GO REGIONS**

---

1 Find the gradients of the lines defined in **a**, **b** and **c**:

a

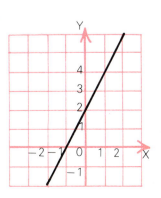

b (i) $y = -3x + 1$
    (ii) $y - x = 2$.

c

x	1	2	3	4
y	2	6	10	14

2 Which of the lines with these equations are parallel?

  a $y = x - 2$ **b** $y = 2x - 1$ **c** $y = -x + 2$ **d** $y = x + 2$ **e** $y = 2x$.

3 Write down the gradient of the line with equation $y = 5x - 1$, and the coordinates of the point where it cuts the y–axis.

4 a Copy and complete the table for $x = -4$ to 2.

b Draw the graph of $y = 2x^2 + 5x - 3$ on 2 mm squared paper.
  Scales: on x–axis, 20 mm to 1 unit
       on y–axis, 20 mm to 10 units.

x	-4	-3	
$2x^2$	32	18	...
$5x$	-20	-15	...
-3	-3	-3	...
y	9	0	...

c Draw the tangents to the curve at the points where:
  (i) $x = -3$ (ii) $x = 0$.
  Calculate their gradients. (Watch the scales).

5 The volume of water (V litres) in a tank is measured after t seconds.

t	0	10	20	30	40	50
V	14	32	42	58	75	89

a Plot the points on 2 mm squared paper.
b Draw the best–fitting straight line.
c Choose two points on the line, and use '$y = mx + c$' to find the equation of the line.
d Calculate the average rate of flow of water into the tank.

6 Describe the unshaded region in each diagram by inequations:

a   **b**   **c**

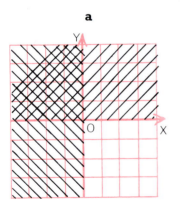 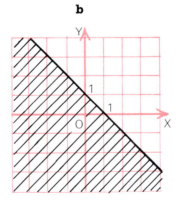 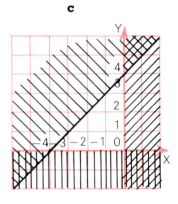

7 a Leave unshaded the region defined by $x \geqslant 0$, $y \geqslant 0$, $x + 2y \leqslant 12$ and $2x + y \leqslant 12$.
  b Find the maximum value of $x + y$ in this region.

DEFINING A SET

Class List
J. Allen
M. Barker
T. Clark
V. Duncan

shopping List
BREAD
ROLLS
CAKES
CREAM

Team List
Jean Bain (capt)
Lorna Jones
Yvonne Brown
Arlene Smith

TOP 10
see List
in record
department

Mathematicians often make lists too—lists of numbers, or points, or shapes, or ideas. For example:

The set of natural numbers, $N = \{1, 2, 3, \ldots\}$.
The set of whole numbers, $W = \{0, 1, 2, 3, \ldots\}$.
The set of integers, $Z = \{\ldots, -3, -2, -1, 0, 1, 2, 3, \ldots\}$.
The set of rational numbers, $Q = \{\text{fractions}\}$, eg $\frac{1}{2}, -\frac{1}{3}, \frac{4}{1}, 3\frac{1}{4}, \ldots$.
The set of real numbers, R, which correspond to all points on the number line, including irrational numbers like $\sqrt{2}$ and π.

They often use symbols too:
$5 \in N$, meaning '5 is a member of set N'.
$-2 \notin W$, meaning '-2 is not a member of set W'.
$n(A) = 4$, meaning 'the number of members, or elements, in set A is 4'.

And set-builder notation:
$A = \{x: 1 < x \leqslant 5, x \in W\}$, 'the set of all x such that x is greater than 1 and less than or equal to 5, where x is a whole number'. $A = \{2, 3, 4, 5\}$.

They insist that a set must be well-defined:
{Clever pupils in your class}—who decides? Not well–defined.
{Even numbers from 10 to 14 inclusive} = $\{10, 12, 14\}$. No doubt; well–defined.

Exercise 1

1A Which of the following are well–defined sets?
    a The set of good-looking pupils in your class.   **b** The set of factors of 12.
    c The set of tall people in your town or village.   **d** The set of multiples of 5.
    e The set of people whose names are listed in the local telephone directory.
    f The set of letters in the word SET.   **g** $\{0, 10, 100, -1\}$.

2A List:
    a the set of natural numbers from 5 to 10 inclusive
    b the set of integers from -1 to 3 inclusive
    c {quadrilaterals which have at least two axes of symmetry}.

3A $A = \{0, 5, 10, 15\}$. List the set of numbers given by:
    a adding 5 to each member of A **b** subtracting 5 from each member of A
    c multiplying each member of A by 5   **d** dividing each member of A by 5.

4A $P = \{a, e, i, o, u\}$, $Q = \{x, y\}$, $R = \{z\}$.

   a Copy and complete the following, using $\in$ or $\notin$:

   (i) $a \ldots P$ (ii) $y \ldots Q$ (iii) $x \ldots P$ (iv) $z \ldots R$

   (v) $u \ldots Q$ (vi) $i \ldots P$ (vii) $o \ldots R$ (viii) $x \ldots Q$.

   b Copy and complete:

   (i) $n(P) = \ldots$ (ii) $n(Q) = \ldots$ (iii) $n(R) = \ldots$.

5A List the following sets. For example, $\{x: 2 \leqslant x \leqslant 6, x \in W\} = \{2, 3, 4, 5, 6\}$

   a $\{x: 1 \leqslant x \leqslant 7, x \in W\}$ **b** $\{y: -2 \leqslant y \leqslant 2, y \in Z\}$

   c $\{t: 4t + 1 = 9, t \in W\}$ **d** $\{p: p^2 = 1, p \in W\}$

   e $\{p: p^2 = 1, p \in Z\}$ **f** $\{(x, y): x + y = 2, x \in W, y \in W\}$.

6A Two dice are rolled.

   a Write down the set P of all possible scores, using:

   (i) the listing method   (ii) set-builder notation.

   b What is $n(P)$?

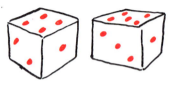

7B Garment sizes are labelled T for tall, S for small.

   a Cathie $\in \{$women of height T cm: $T \geqslant 170\}$ Sheila $\in \{$women of height S cm: $S \leqslant 160\}$. What can you say about garment sizes for: (i) Cathie   (ii) Sheila?

   b Describe: (i) in words   (ii) using set-builder notation, the medium height category M between T and S.

8B $V = \{$vertices$\}$ and $S = \{$sides$\}$ of $\triangle$ABC.

   a (i) List the members of V and S (ii) Write down $n(V)$ and $n(S)$.

   b Repeat **a** for: (i) quadrilateral ABCD   (ii) pentagon ABCDE.

   c Can you make a general statement about $n(V)$ and $n(S)$?

9B Given a set of data, and other information, a computer can search a list of addresses or telephone numbers, increase everyone's wage by 10%, and so on. Read through this program, checking that it takes $A = \{2, 3, 5, 7, 11, 13\}$, and generates $\{4, 5, 7, 9, 13, 15\}$.

   1∅   FOR X = 1 TO 6 .... $n(A) = 6$

   2∅   READ a .... $a \in A$

   3∅ $b = a + 2$ .... the *generating rule*

   4∅   PRINT b

   5∅   NEXT X

   6∅   DATA 2, 3, 5, 7, 11, 13 .... A.

   a Write down $n(A)$ and the generating rule for:

   (i) $A\{2, 3, 4\} \rightarrow B\{4, 6, 8\}$ (ii) $A = \{5, 10, 15, 20\} \rightarrow B\{-5, -10, -15, -20\}$.

   b Write down $n(A)$ and the output set B for:

   (i) $A = \{0, 3, 6, 9\}$, generating rule $b = a - 3$ (ii) $A = \{50, 100\}$, generating rule $b = \frac{1}{5}a$.

   c Choose a set A and generating rule of your own. Write out the program, and list the output set B. Run your program in a computer, if possible.

EQUAL SETS

These two sets of cards are *equal*. They contain exactly the same members, or elements. The order does not matter.

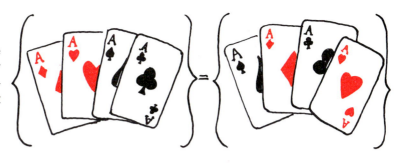

> Two sets A and B are equal if they contain exactly the same elements.

=== *Exercise 2* ===

1A In which of these pairs are the two sets equal?
    a $P = \{a, b, c\}$ and $Q = \{c, b, a\}$ **b** $R \{0, 2, 4, 6\}$, $S = \{4, 2, 6\}$
    c $M = \{-4, -3, -2, -1\}$ and $N = \{x: -4 \leqslant x \leqslant -1, x \text{ an integer}\}$.

2A $\{1, 2, 3\} = \{2, 1, 3\} = \ldots$ Can you write down four more equal sets?

Members of a set need not all be different. Although $12 = 2 \times 2 \times 3$, the set of prime factors of 12 is usually written $\{2, 3\}$.

3A List the set of prime factors of:
    a 4   **b** 9   **c** 20   **d** 32   **e** 36   **f** 45.

4A List the 'solution set' of each equation, where x is an integer:
    a $(x-2)(x+2) = 0$ **b** $x(x-5) = 0$ **c** $(x-1)(x-1) = 0$.

5A List the set of keys required to type the words:
    a ASSESS   **b** MISSISSIPPI   **c** REPETITION.

6A List the set of stencils needed to print:
    a COMMITTEE ROOM
    b SUPERCALIFRAGILISTICEXPIALIDOCIOUS.

7A Paul maintains that $A = B \Rightarrow n(A) = n(B)$, *and* $n(C) = n(D) \Rightarrow C = D$. His friend Stan disagrees. Who is correct? Explain.

8A As order does not matter, {M, O, R, A, G} = {M, A, R, G, O}. Can you think of other names of people or towns like this?

* * * * *

The Golf Club has closed (no members), the stamp album is empty (no stamps), the larder is bare (no food). All are examples of **the empty set**—the set with no members.
In mathematics its symbol is { } or $\varnothing$. For example, $\{x: 4x+1 = 6, x \in W\} = \varnothing$. Why?

9A Which of the following are examples of the empty set?
 a {cities in Britain with a population of more than a million}
 b {pupils in your class who travel to school by helicopter}
 c {prime numbers less than 5} **d** {negative numbers greater than zero}
 e {points which lie both on the x–axis *and* the y–axis}.

10A x stands for a whole number or a fraction. Which equations have the empty set as their solution set?
 a $2x+1 = 5$ **b** $3x-1 = 4$ **c** $4x+8 = 0$ **d** $x^2 = 1$
 e $x^2 = -1$ **f** $x^3 = -1$ **g** $\sin x° = 2$ **h** $2\cos x° = 1$.

11A Your calculator is clever at coping with the empty set.

 a Try these: (i) | 1 | | ÷ | | 0 | | = | (ii) | 1 | | +/– | | √ | (iii) | 2 | | 2ndF | | sin |

 b Explain why each is the empty set.
 c Try to find some more examples with your calculator.

12A Only acids turn blue litmus paper red. What can you say about the set of alkalis that turn it red?

13A Make up three or four examples of the empty set.

---

 1 Find a rule, or property, which separates the members of your class into two sets.

 2 Compare your rule, or property, with your neighbour's.

 3 Make up a set of rules, or properties, which would identify any one particular pupil in the school.

SUBSETS—SETS WITHIN SETS

To help her in her driving test, Jenny arranged this set R of road signs into three **subsets**:

$R = $

| 1 | 2 | 3 | 4 | 5 | 6 | 7 | 8 | 9 | 10 |

$W = \{$signs giving a warning$\}$, $Y = \{$signs saying 'You must . . .'$\}$,
$N = \{$signs saying 'You must not . . .'$\}$.
The set R contains *subsets* W, Y and N.
In symbols, $W \subset R$, $Y \subset R$ and $N \subset R$.

In a Venn diagram,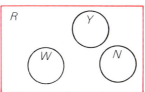

John Venn was an English mathematician who lived from 1834 to 1923

> A set A is a subset of set B if every member of A is a member of B.

Exercise 3

1A a List the subsets W, Y and N of road signs above, using the numbers below them.
  b What do you notice about the shapes of the outlines of the signs in each subset?

2A

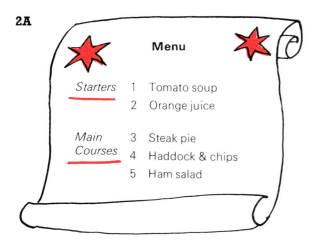

The menu has a set of five items.
a List all possible subsets for a 2-course meal (starter and main course), for example $\{1, 5\}$.
b How many different 2-course meals could you choose?

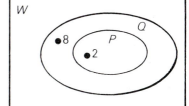

3A $W = \{$whole numbers$\}$, $P = \{2, 4, 6\}$ and $Q = \{2, 4, 6, 8, 10\}$.
  a Link pairs of W, P, Q using the symbol $\subset$.
  b Copy and complete the Venn diagram.
    Mark each element with a dot and its 'name', like the two that are shown.

4A S = {factors of 4}, T = {factors of 6} and V = {factors of 12}.
  a List S, T and V.
  b Link pairs of the sets, using the symbol $\subset$.

5A a Put $\subset$ and $=$ between as many pairs as you can:
    A = {0, 2, 4}, B = {0}, C = {2, 4}, D = {4, 2}.
  b Which of these are true and which are false? Remember that $A \subset B$ means A is a subset of
    B, so A is part of, or equal to B.
    (i) $\{4, 5\} \subset \{5, 2, 4\}$ (ii) $\{5\} \subset \{5\}$ (iii) $\{5, 4\} \subset \{5, 2\}$

(iv)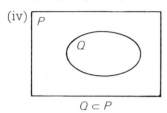
$Q \subset P$

(v)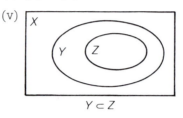
$Y \subset Z$

(vi)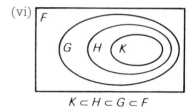
$K \subset H \subset G \subset F$

6A Draw Venn diagrams to illustrate the relationship between the sets in each pair:
  a M = {2, 4, 6}, N = {4}   **b** P = {−1, 0, 1}, Q = {integers}
  c T = {triangles}, I = {isosceles triangles}   **d** P = {parallelograms}, R = {rectangles}.

7B The subsets of $\{x\}$ are $\{x\}$ and $\{\ \}$ or $\varnothing$. List the *four* subsets of $\{x, y\}$.

8B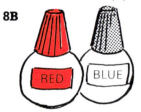

Amanda is adding colouring to her cake icing. From C = {red, blue}
she can have four colour subsets: {red}, {blue}, {red, blue}, { }.
  a What colour would the last one give?
  b List all the subsets of colours she can have from three bottles of
    colouring—red, blue and yellow.
  c How many colours could she have with only one bottle?
  d Copy and complete:

Number of members in set	1	2	3	4	5	...	n
Number of subsets						...	

  e Use the $\boxed{y^x}$ key on your calculator to find the number of subsets for (i) 10 (ii) 100,
  members in the set.

9B How many different little blocks of
paint must be in Superchoice's Paint
tins to avoid difficulty with the Trades
Description Act?

Over 1 000 000 different colours!

a Investigate the sets of weighings you could make with only the: (i) 1 g weight
          (ii) 1 and 2 g weights
          (iii) 1, 2 and 4 g weights
          (iv) 1, 2, 4 and 8 g weights
          (v) complete set of weights.

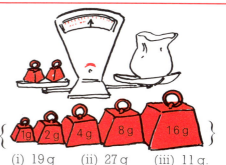

Can you find a formula for n weights? $W = \{$ 1g, 2 g, 4 g, 8 g, 16 g $\}$

b List the subsets of weights needed to weigh: (i) 19 g   (ii) 27 g   (iii) 11 g.
Is there only one possible subset for each?

THE UNIVERSAL SET

A universal set contains all the elements being discussed.

For example, a universal set for $\{2, 4, 6\}$ could be:
{even numbers}, or {whole numbers}, or {integers}, or {numbers less than 10}.
The symbol for a universal set is E.

Example
$E = \{$colours of the rainbow$\}$, and
$C = \{$red, orange, yellow$\}$
are shown in the Venn diagram.

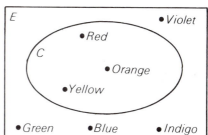

Exercise 4

1A Write down a possible universal set for:
    a $\{1, 3, 5\}$ **b** $\{x, y, z\}$ **c** {pupils in your class}.

2A Draw a Venn diagram for $E = \{$negative numbers$\}$, $N = \{-10, -100, -1000\}$.

3A Tina empties her purse, and writes down $\{5p, 10p, 50p\}$. What could be her universal set?
Try to think of several possibilities.

4A The universal set is $E = \{2, 3, 4, 5, 6, \ldots, 12\}$. List these subsets:
    a (i) $V = \{$even numbers$\}$ (ii) $P = \{$prime numbers$\}$.
    b The solution sets of: (i) $2x - 3 > 9$ (ii) $(x-3)(x+4) = 0$ (iii) $5x + 4 = 2x + 1$.

5A Copy and complete this Venn diagram for the sets E, V and P in question **4A a**.

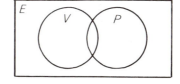

6B Choose a universal set and illustrate in a Venn diagram {planets nearer the sun than the earth} and {planets further from the sun than the earth}.

Diversion
What is this all about?

Use the definitions of N, W, Z, Q and R on page 197 to draw a Venn diagram to illustrate the relations between these sets. Take R as the universal set. Mark a few elements in each set.

INTERSECTION OF SETS

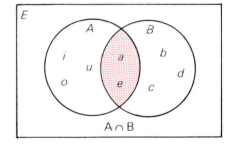

$E = \{$letters of the English alphabet$\}$, $A = \{$vowels$\}$ and $B = \{$first five letters of the alphabet$\}$.

$A = \{a, e, i, o, u\}$ and $B = \{a, b, c, d, e\}$.

The members of $\{a, e\}$ belong to set A *and* to set B.

$\{a, e\}$ is the *intersection* of sets A and B. In symbols, $A \cap B = \{a, e\}$.

The intersection of two sets A and B is the set whose elements are in both A and B.

=== *Exercise 5* ===

1A $A = \{1, 3, 5, 6, 8\}$, $B = \{4, 6, 8\}$, $C = \{1, 3, 8\}$ and $D = \{6\}$.
List: **a** $A \cap B$ **b** $A \cap C$ **c** $B \cap C$ **d** $C \cap D$.

2A List $A \cap B$ for each Venn diagram.

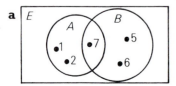
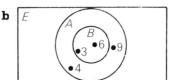
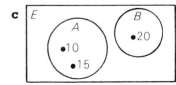

In **c**, A and B have no elements in common. They are called *disjoint* sets.

THE MEANING AND USE OF A SET

3A E = {teachers in your school},
   T = {your own teachers} and
   M = {mathematics teachers in the school}.
   Describe the set $T \cap M$.

4A E = {whole numbers from 1 to 100 inclusive}, S = {squares of whole numbers from 1 to 10} and C = {cubes of whole numbers from 1 to 4}.
   a List S, C and $S \cap C$.
   b Draw a Venn diagram.

5A E = {1, 2, 3, ..., 12}, P = {multiples of 2} and Q = {multiples of 3} in E.
   a List P, Q, $P \cap Q$, and draw a Venn diagram.
   b What is the connection between $P \cap Q$ and the lcm of 2 and 3?

6A Repeat question **5A** for E = {1, 2, 3, ..., 24}, Q = {multiples of 3} and R = {multiples of 4}.

7A E = {−3, −2, −1, 0, 1, 2, 3, 4, 5}. List these sets:
   a A = {x: $2x > -4$}   **b** B = {x: $x + 1 < 4$}   **c** $A \cap B$.

8A Describe in words:
   a {colours in the rainbow} $\cap$ {colours in the Union Jack flag}
   b {quadrilaterals with all their sides equal} $\cap$ {quadrilaterals with all their angles equal}.

9B *An evening's viewing*

BBC1	BBC2	ITN	CHANNEL 4
7.00 Holiday	6.50 Rally	7.00 Emmerdale	7.00 News
7.30 Westender	7.45 Smash	7.30 News	7.50 Comment
8.00 Sorry	8.50 Food & drink	8.00 Comedians	8.00 Brookend
8.30 Watch	9.00 Film	9.00 Boon	8.30 Money
9.00 News	10.35 Newsnight	10.00 News	9.30 Film

a You can see 'Sorry' at times T = {times from 8.00 to 8.30} on BBC1,
      and 'Smash' at times R = {times from 7.45 to 8.50} on BBC2.
   (i) Describe $T \cap R$ as a set of times.
   (ii) Why would some viewers prefer $T \cap R = \varnothing$?
b In the same way, describe:
   (i) A = set of times for ITN early news
       B = set of times for Ch4 news and $A \cap B$.
   (ii) C = times for 'Watch', D = times for 'Boon' and $C \cap D$.

$A = \{1, 3, 5, 7, 9, 11, 13, 15\}$
$B = \{2, 3, 6, 7, 10, 11, 14, 15\}$
$C = \{4, 5, 6, 7, 12, 13, 14, 15\}$
$D = \{8, 9, 10, 11, 12, 13, 14, 15\}.$

a List:

(i) $A \cap B$

(ii) the least member of $A \cap B$

(iii) the sum of the least members of A and B.

b Repeat **a** for $A \cap B \cap C$ and for $A \cap B \cap C \cap D$.

c A party cracker novelty consists of 5 cards, each with a set of numbers on it. You ask someone to think of a number from 1 to 32 and to say which of the cards contain the number. You should be able to say what the number is, almost immediately. Design the set of cards. (Look carefully at the way in which cards A–D are built up, one number at a time. Add card E, extending the sets of numbers in A–D as you go.)

IT'S MAGIC

Practical

Draw three rectangles 6 cm by 3 cm on tracing paper. Draw and colour the flag of St Patrick on one, St Andrew on another and St George on the third.

Cut out the flags and place them on top of each other to obtain the Union Jack. Now you're ready for the union of sets.

UNION OF SETS

$E = \{$all the tools in the car$\}$.

To take a wheel off, you'll need:

$W = \{$jack, wheel-brace, screw driver$\}$.

To adjust the brakes, you'll need:

$B = \{$jack, wheel-brace, spanner$\}$.

To do both jobs, combine W and B:

$\{$jack, wheel-brace, screw driver, spanner$\}$.

This combined set is called the *union* of W and B, written $W \cup B$. (Compare $\cup$ and **U**nion).

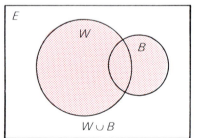

E

W

B

$W \cup B$

The union of two sets A and B is the set of elements in A or B, or in A and B.

1A $A = \{p, q, r\}$ and $B = \{r, s, t\}$. List $A \cup B$.

2A List the members of $A \cup B$ for each diagram:

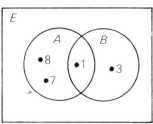

a

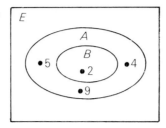

b

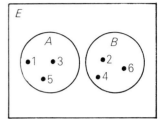

c

3A Copy these Venn diagrams, and shade the region described:

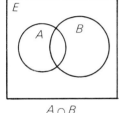

a

$A \cap B$

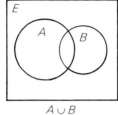

b

$A \cup B$

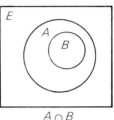

c

$A \cap B$

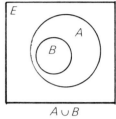

d

$A \cup B$

4A

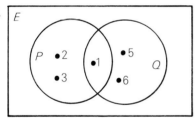

    a List: (i) P (ii) Q (iii) $P \cap Q$ (iv) $P \cup Q$.
    b Check that $n(P \cup Q) + n(P \cap Q) = n(P) + n(Q)$.

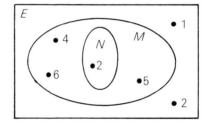

5A a List: (i) $M \cup N$ (ii) $M \cap N$.
    b Check that $n(M \cup N) = n(M)$.

6A

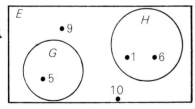

    a List: (i) $G \cup H$ (ii) $G \cap H$.
    b Check that:
      (i) $n(G \cup H) = n(G) + n(H)$ (ii) $n(G \cap H) = 0$.

7B Birth to death in set notation:
   Robert the Bruce's lifetime, $B = \{y: 1274 \leqslant y \leqslant 1329, y$ *a whole number*$\}$.
   His son David II's lifetime, $D = \{y: 1324 \leqslant y \leqslant 1371, y$ *a whole number*$\}$.
   a Write $B \cup D$ and $B \cap D$ in set–builder notation.
   b Describe in words what each means.

8B Draw two Venn diagrams with two intersecting sets. Fill in *numbers* to solve these
   problems:
   a $n(A) = 14$, $n(B) = 26$ and $n(A \cap B) = 11$. Find $n(A \cup B)$.
   b $n(C) = 9$, $n(D) = 10$ and $n(C \cup D) = 15$. Find $n(C \cap D)$.

COMPLEMENT OF A SET

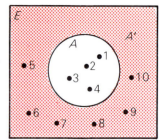

$E = \{1, 2, 3, \ldots, 10\}$ and $A = \{1, 2, 3, 4\}$.

The set of elements in E, but *not* in A, is called the *complement* of A.
In symbols, $A' = \{5, 6, 7, 8, 9, 10\}$.

THE MEANING AND USE OF A SET

> The complement of a subset A of a universal set E is the set of elements of E which do not belong to A. In symbols, $A' = \{x : x \notin A\}$.

=== *Exercise 7* ===

1A $E = \{0, 1, 2, 3, 4, 5\}$ and $A = \{2, 4\}$.
   a List A'.   **b** Illustrate in a Venn diagram.

2A $W = \{\text{whole numbers}\}$, $V = \{\text{even numbers}\}$. Describe the complement of V.

3A Copy the Venn diagrams, and shade the regions described:

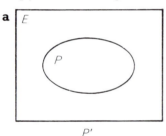

P'

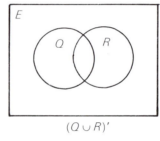

$(Q \cup R)'$

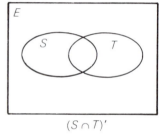

$(S \cap T)'$

4B A geologist's cross–section of the earth. $E = \{h : -10 \leqslant h \leqslant 8, h \text{ any number}\}$.

a Name the part of the cross–section described by $T = \{h : 0 \leqslant h \leqslant 8\}$.

b (i) Write down T' in set notation.
    (ii) Describe it in the diagram.

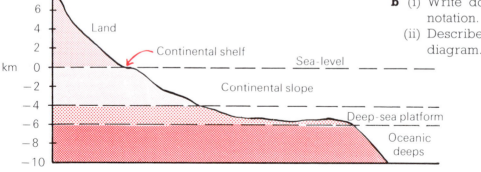

c Name the part of the cross–section described by $P = \{h : -10 \leqslant h \leqslant -6\}$.

d (i) Write down P' in set notation.   (ii) Describe it in the diagram.

5B $A \subset E$. Copy and complete: **a** $A \cup A' = \ldots$ **b** $A \cap A' = \ldots$.

6B $E = \{x: 0 \leqslant x \leqslant 6, x \text{ a whole number}\}$ and $A = \{x: x > 4\}$.
Write down A', and illustrate in a Venn diagram.

7B Describe the sets shown by the shaded regions in terms of unions, intersections or complements of sets A, B and C.

a 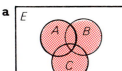 **b** 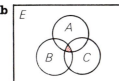 **c** 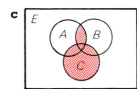 **d**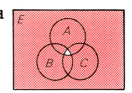

a In arithmetic and algebra,
  (i) $a \times (b + c) = (a \times b) + (a \times c)$, but (ii) $a + (b \times c) \neq (a + b) \times (a + c)$ in general.
  Check these by replacing a, b and c by suitable numbers of your choice.

b Georges Boole (1815–1864) invented a kind of algebra which used sets instead of numbers, and the operations of union and intersection instead of multiplication and addition. Use shading in Venn diagrams to investigate whether or not for sets A, B and C:
  (i) $A \cap (B \cup C) = (A \cap B) \cup (A \cap C)$ (ii) $A \cup (B \cap C) = (A \cup B) \cap (A \cup C)$.

VENN DIAGRAMS AS MATHEMATICAL MODELS

Exercise 8

1A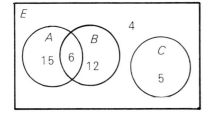

E is the set of people living in East Avenue.
$A = \{\text{readers of the 'Daily Moan'}\}$. $n(A) = 21$.
$B = \{\text{readers of the 'Daily Groan'}\}$. $n(B) = 18$.
$C = \{\text{readers of the 'Daily Cheer'}\}$. $n(C) = 5$.

a How many read:
  (i) both the 'Daily Moan' and the 'Daily Groan'
  (ii) the 'Daily Groan' only
  (iii) only one of the three newspapers?
b How many people live in East Avenue?

2A $E = \{\text{members of Action Youth Club}\}$
  $T = \{\text{members in the table–tennis team}\}$
  $D = \{\text{members in the darts team}\}$
  $S = \{\text{members in the snooker team}\}$.
  a How many members are there in the Youth Club?
  b How many play in:

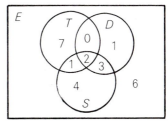

  (i) all three teams   (ii) the table–tennis team   (iii) both darts and snooker teams
  (iv) the table–tennis and snooker teams but not in the darts team?

3A

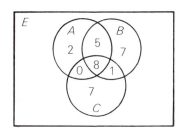

$E = \{$The 92 football clubs in the 4 English Divisions$\}$
$A = \{$League Championship winners 1946–86$\}$
$B = \{$League Cup winners, 1946–86$\}$
$C = \{$FA Cup winners, 1946–86$\}$.
How many teams have won:

a all three
b FA cup only
c League Championship and League Cup
d none of them?

4A A and B are subsets of a universal set E. $n(E) = 44$, $n(A) = 25$, $n(B) = 22$ and $n(A \cap B) = 8$. Draw a Venn diagram, and calculate:
a $n(A \cup B)$ **b** $n(A \cup B)'$. Describe **b** in words.

5A

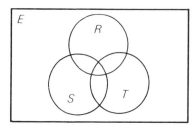

Copy the Venn diagram and fill in the number of elements in each region:
$n(R \cap S \cap T) = 6$, $n(R \cap S) = 7$, $n(S \cap T) = 9$,
$n(R \cap T) = 11$, $n(R) = 14$, $n(S) = 13$, $n(T) = 22$
and $n(R \cup S \cup T)' = 4$. Calculate $n(E)$.

6A A group of teenagers were asked whether they had ever had mumps (M), measles (S) or chickenpox (C). How many had caught:
a all three illnesses
b only two of the illnesses
c only one of the illnesses
d none of them?

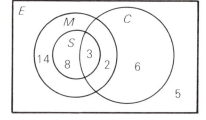

7A Divot Golf Club has 180 members. 105 play most Saturdays, 95 most Sundays and 45 most Saturdays and Sundays. Draw a Venn diagram, and find how many members don't usually play at weekends.

8B Half the families in Cherry Tree Drive went abroad on holiday last year—35 during the summer, 18 during the winter and 12 both in summer and in winter. How many families live in the Drive?

9B 55 pupils were asked if they watched darts or snooker on TV: 18 replied 'Darts', 35 'Snooker' and 13 'Neither'.

  a Draw a Venn diagram, with x for the number who watched both darts and snooker.

  b Make an equation and solve it to find x.

10B A sample of people were asked for their opinion on the main causes of heart illnesses: 2 people blamed three causes—alcohol, diet and smoking. Others named two causes only—4, smoking and alcohol; 9, alcohol and diet; 10, smoking and diet. Others named only one cause—12, smoking; 7, alcohol and 8, diet.

  a How many people were in the sample?

  b Which of the three factors was considered most important?

11B 49 fourth year students took tests in French, German and Latin:

25 passed in French, 22 in German, 19 in Latin, 4 in French and German but not Latin, 6 in French and Latin but not German, and 2 in German and Latin but not French—9 failed in all three subjects. Calculate the number who passed in:

  a all three subjects   **b** exactly two subjects   **c** just one subject.

---

1 Gather data on some of the following, using Venn diagrams to display and analyse the data:

  a Pupils in your class who are in sports teams, or have favourite TV programmes, or have chosen different sets of school subjects.

  b The continents and oceans of the world and their situation—in the Northern or Southern hemisphere, or in both.

  c Possible scores from doubles and trebles at darts.

  d Track, field and decathlon events in the Olympic Games.

2 Scientists and others who use and apply mathematics often classify data by means of sets, subsets and intersections of sets. Investigate some of the following:

  a The Periodic Table of chemical elements.

  b The classification of plants

  c Methods of classifying earthquakes, windspeeds, hardness of metals, temperature of hot steel.

1 $A = \{-2, 0, 2, 4, 6\}$. List the set of numbers given by:
  a adding 4 to each member of A **b** subtracting 4 from each member of A
  c multiplying each member of A by: (i) 2   (ii) -2 (iii) 0.

2 List: **a** $\{x : 4 \leqslant x < 10, x \in W\}$
        b the set of quadrilaterals which have only two axes of symmetry.

3 $A = \{p, q, m, n\}$ and $B = \{q, r, s\}$. Copy and complete:
  a $p \ldots A$ **b** $p \ldots B$ **c** $\ldots \in A$ and B **d** $n(A) + n(B) = \ldots$

4 $P = \{0, 2, -1\}$, $Q = \{2, 1, -1\}$, $R = \{-1, 2\}$, $S = \{-1, 0, 2\}$ and $T = \{-1, 1, 2\}$.
  a Link pairs of equal sets.   **b** Which elements are in all of the sets?

5 a Describe and list two subsets of coins from the ones shown below. Call them A and B.
  b Say whether or not the 2p coin and the 20p coin are members of A and B. (Use $\in, \notin$.)

6 a $E = \{$real numbers$\}$. Write down the solution set of:
    (i) $6x + 1 = 10$ (ii) $(2x + 1)(2x + 1) = 0$ (iii) $2x - 3 = 3x - 2$.
  b List the set of keys needed to type ANTIDISESTABLISHMENTARIANISM.

7 $E = \{20, 21, 22, \ldots, 30\}$. List the subsets consisting of:
  a multiples of 3   **b** prime numbers   **c** factors of 60.

8 $A = \{2, 4, 6, 8\}$, $B = \{2, 8\}$, $C = \{8, 10\}$ and $D = \{3\}$. Draw Venn diagrams to illustrate:
  a $A \cap B$ **b** $B \cap C$ **c** $C \cap D$.

9 List: **a** $P \cap Q$ **b** $P \cup Q$ **c** $P \cap R$
        d $P \cap Q \cap R$ **e** $P \cup Q \cup R$.

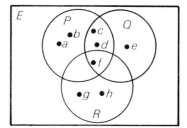

10 $E = \{$quadrilaterals$\}$, $R = \{$rectangles$\}$, $B = \{$rhombuses$\}$.
  Describe: **a** $R \cap B$ **b** $R \cup B$ **c** $(R \cup B)'$.

11 $E = \{$whole numbers$\}$. For which equation(s) is the solution the empty set?
  a $x + 1 = 0$ **b** $2x - 3 = 0$ **c** $x^2 - 4 = 0$ **d** $3(2x - 1) + 5 = 8$.

Class discussion

1 a How many pupils are in the class?
 b How much money have you with you?
 c What time is it?
 d Estimate the room temperature.
 e What is the width of this page?

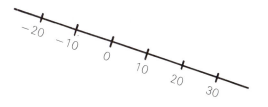

All of these quantities have one thing in common—size, or *magnitude*. The magnitude of each can be shown on the number scale. So they are called *scalars*.

2 Concorde takes off from Heathrow, and flies in a straight line for 60 km.
 a Why are you unable to say exactly where it is? Its course is 090°, due East.
 b Can you now fix its position?

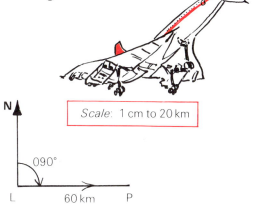

Scale: 1 cm to 20 km

'60 km due East' involves distance, or *magnitude*, and *direction*.

A quantity which has magnitude and direction is a **vector**.

One example of a vector is a *displacement*, that is the distance moved by a point in a certain direction. Concorde's displacement of 60 km due East is represented in *distance and direction* by the arrowed line drawn *from* L *to* P above, 3 cm long.

=========================== *Exercise 1* ===========================

1

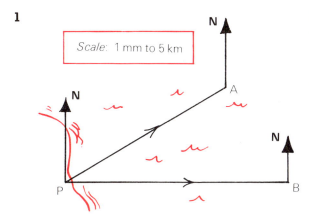

Scale: 1 mm to 5 km

Helicopter C for Charlie flies from its service port P to oilrigs A and B. By measuring both distance and direction, describe the helicopter's displacement in a flight from:
a P to A **b** P to B.

2 Bill is batting for the 1st XI. In one over he hit the cricket ball to points 1, 2, 3 and 4. Describe the displacement of each ball by giving its distance from the bat B, and its direction by the angle its line makes with the line BW between the wickets.

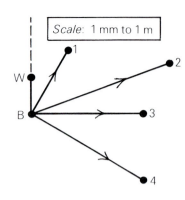

Scale: 1 mm to 1 m

3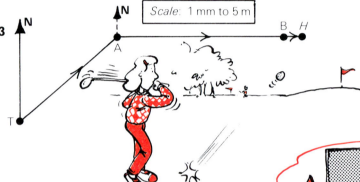

Scale: 1 mm to 5 m

Susan scored a 'birdie' 3 at the fourteenth hole. Describe the displacement of the golf ball from:

a T to A   **b** A to B
c B to H.

4 Alan takes the lift from the ground floor to the fourth floor. Each storey is 3 m high. He then turns right and walks 4 m to his flat. Draw a 'vector diagram' of his journey, using a scale of 1 cm to 2 m.

VECTORS AND DIRECTED LINE SEGMENTS

Vectors are named in black type **u**, **v**, **w**, ... You should write them underlined — u̲ or ṵ, ...

A vector can be represented in magnitude and direction by a *directed line segment* $\overrightarrow{AB}$.

The length of $\overrightarrow{AB}$ is proportional to the magnitude of **u**, and the arrow shows the direction of **u**.

In the diagram, $\overrightarrow{AB}$ and $\overrightarrow{CD}$ have the same magnitude and direction, so $\overrightarrow{AB} = \overrightarrow{CD}$.

Both represent the vector **u**. $\boxed{u = \overrightarrow{AB}.}$

1 Name the directed line segment equal to:

a $\overrightarrow{AB}$ **b** $\overrightarrow{BC}$ **c** $\overrightarrow{CD}$ **d** $\overrightarrow{AC}$.

Did you remember to check directions as well as lengths?

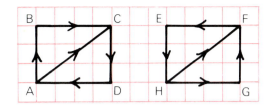

2

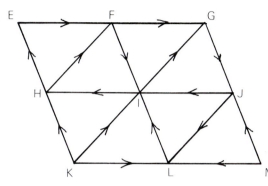

The directed line segments make a tiling of congruent parallelograms.

a Name all the directed line segments equal to:

(i) $\overrightarrow{EF}$ (ii) $\overrightarrow{HE}$ (iii) $\overrightarrow{HF}$ (iv) $\overrightarrow{FI}$.

b Name a directed line segment with the same magnitude, but opposite direction to:

(i) $\overrightarrow{EF}$ (ii) $\overrightarrow{EH}$ (iii) $\overrightarrow{EG}$.

c Name a directed line segment with double the magnitude and the same direction as:

(i) $\overrightarrow{EF}$ (ii) $\overrightarrow{HF}$ (iii) $\overrightarrow{LI}$.

3 Name all the directed line segments representing vectors:

a u **b** v **c** w

4 a Plot the points A(1, 3), B(3, 4), C(2, 2).

b $\overrightarrow{AB}$ and $\overrightarrow{CD}$ both represent vector p. Find the coordinates of D.

c $\overrightarrow{AC}$ and $\overrightarrow{EB}$ both represent vector q. Find the coordinates of E.

5

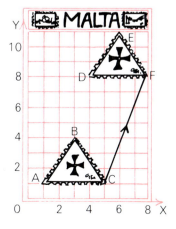

Bryan is rearranging his stamps.

a Draw △ABC on squared paper. C is the point (5, 1). He slides his stamp to position DEF. F is (8, 8).

b Draw △DEF. Draw in dotted lines, or in colour, the directed line segments $\overrightarrow{AD}$, $\overrightarrow{BE}$ and $\overrightarrow{CF}$ which represent the displacement of his stamp. What can you say about these segments?

COMPONENTS

The displacement from A to B can be made by moving 2 units in the x–direction, then 4 units in the y–direction.

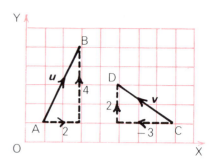

> The vector $\overrightarrow{AB} = \begin{pmatrix} 2 \\ 4 \end{pmatrix}$, or $\boldsymbol{u} = \begin{pmatrix} 2 \\ 4 \end{pmatrix}$.

2 is the x–component of $\overrightarrow{AB}$, and 4 is the y–component.

In the same way, $\overrightarrow{CD} = \begin{pmatrix} -3 \\ 2 \end{pmatrix}$, or $\boldsymbol{v} = \begin{pmatrix} -3 \\ 2 \end{pmatrix}$.

Exercise 3

1 Write each vector in component form like this: $\boldsymbol{u} = \begin{pmatrix} 2 \\ 3 \end{pmatrix}$.

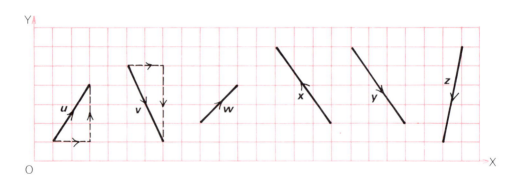

2 a Draw a directed line segment on squared paper to represent each vector:

(i) $\boldsymbol{p} = \begin{pmatrix} 5 \\ 2 \end{pmatrix}$ (ii) $\boldsymbol{q} = \begin{pmatrix} -5 \\ -2 \end{pmatrix}$ (iii) $\boldsymbol{r} = \begin{pmatrix} 6 \\ -1 \end{pmatrix}$ (iv) $\boldsymbol{s} = \begin{pmatrix} -3 \\ 3 \end{pmatrix}$.

b Draw two more directed line segments for $\begin{pmatrix} 5 \\ 2 \end{pmatrix}$. What can you say about:

(i) their length   (ii) their direction?

3 $\overrightarrow{PQ} = \begin{pmatrix} 3 \\ 2 \end{pmatrix}$. Find the coordinates of Q, given that P is the point:

a $(0, 0)$ **b** $(1, 4)$ **c** $(-4, 3)$ **d** $(5, -2)$.

4 In chess, the Knight can move 2 squares left or right then 1 square up or down, *or* 1 square left or right then 2 squares up or down. Write down the components for his moves to squares A, B, ..., H. For example, $\overrightarrow{KG} = \begin{pmatrix} -2 \\ 1 \end{pmatrix}$.

5 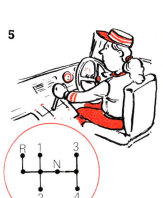 Irma drives a bus. She selects gears by sliding the gear–lever along the grooves. The distance between the gear positions (shown by the dots) is 3 cm.

a Use components, with N as origin, to describe the displacement of the gear lever from:

(i) N to 3 (ii) N to 4 (iii) N to 1 (iv) 1 to 2 (v) 3 to 4.

b Express: (i) $\overrightarrow{NR}$ (ii) $\overrightarrow{RN}$ in components. Comment.

6 $O(0, 0)$, $P(2, 1)$, $Q(-4, 3)$, $R(-3, -2)$, $S(x, y)$ undergo the displacement $\begin{pmatrix} 2 \\ 3 \end{pmatrix}$.

a Write down the coordinates of the new positions of the points.

b Write down the image of the point (x, y) under the displacement $\begin{pmatrix} a \\ b \end{pmatrix}$.

THE MAGNITUDE OF A VECTOR

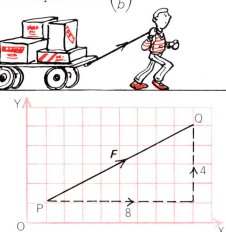

Pat is pulling a trolley across the floor. He pulls with a *force* **F** which has size and direction. So force is another example of a vector.

The horizontal component of the force is 8 newtons, and the vertical component is 4 newtons.

$\overrightarrow{PQ} = \begin{pmatrix} 8 \\ 4 \end{pmatrix}$, and the length of PQ is written $|\overrightarrow{PQ}|$.

By Pythagoras' Theorem, $|\overrightarrow{PQ}|^2 = 8^2 + 4^2$,

and $|\overrightarrow{PQ}| = 8.9$, correct to 1 decimal place.

So Pat pulls the trolley with a force **F** of magnitude 8·9 newtons.

> The magnitude of vector **u** is written $|\boldsymbol{u}|$.
>
> If $\boldsymbol{u} = \begin{pmatrix} -9 \\ 12 \end{pmatrix}$, $|\boldsymbol{u}|^2 = (-9)^2 + (12)^2 = 225$. So $|\boldsymbol{u}| = 15$.

Exercise 4

1 Draw each vector on squared paper, then calculate its magnitude. Write your answer like this: $|\underset{\sim}{y}| = 6$.

a $\boldsymbol{u} = \begin{pmatrix} 4 \\ 3 \end{pmatrix}$ **b** $\boldsymbol{v} = \begin{pmatrix} 6 \\ 8 \end{pmatrix}$ **c** $\boldsymbol{w} = \begin{pmatrix} 5 \\ -12 \end{pmatrix}$ **d** $\boldsymbol{x} = \begin{pmatrix} -3 \\ 4 \end{pmatrix}$.

2 a Write down the components of the forces represented by $\overrightarrow{AB}$.
  b Calculate the magnitude of each force, correct to 1 decimal place where necessary.

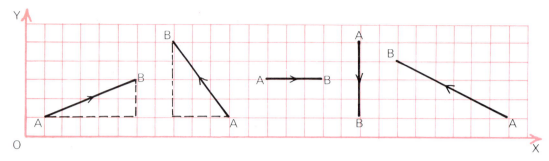

3

An orienteering course requires a distance and a direction, so is yet another example of a vector. Calculate the lengths of the courses, to the nearest km, given by these vectors:

a $\overrightarrow{PQ} = \begin{pmatrix} 5 \\ 5 \end{pmatrix}$ **b** $\overrightarrow{RS} = \begin{pmatrix} 1 \\ 9 \end{pmatrix}$ **c** $\overrightarrow{TU} = \begin{pmatrix} 6 \\ 4 \end{pmatrix}$.

4 A velocity has a magnitude (speed) and a direction. So it is a vector also. Calculate the magnitude, to the nearest m/s where necessary, of the velocities given by these vectors:

a $\boldsymbol{v} = \begin{pmatrix} 7 \\ 24 \end{pmatrix}$ **b** $\boldsymbol{v} = \begin{pmatrix} 12 \\ -9 \end{pmatrix}$ **c** $\boldsymbol{v} = \begin{pmatrix} 10 \\ 10 \end{pmatrix}$.

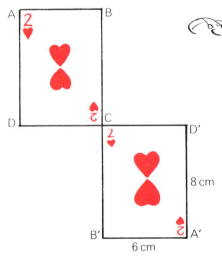

5

Under a half-turn about C, the card ABCD moves to the position A'B'CD'.
  a Write down the magnitude and direction (with reference to AB) of:
  (i) $\overrightarrow{BB'}$ (ii) $\overrightarrow{DD'}$ (iii) $\overrightarrow{AA'}$.

  b Card ABCD can slide perfectly to CD'A'B' under the displacement $\overrightarrow{AC}$.
    (i) Name three other directed line segments for this displacement.
    (ii) Write down the magnitude of the displacement.

ADDITION OF VECTORS

Adding *scalars* is easy; for example, $15 + 7$, £8 + £7, $6° + 9°$, 14 m + 8 m.
Adding *vectors* is not so obvious, because direction as well as magnitude is involved. Here are two ways to do it—using directed line segments and using components.

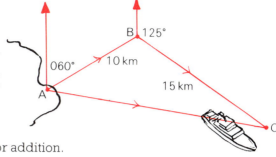

1 The Silver Sea leaves port A on a bearing of 060°, and sails in a straight line for 10 km to B. She then changes course to 125° for 15 km to C.
$\overrightarrow{AB}$ followed by $\overrightarrow{BC}$ results in $\overrightarrow{AC}$.

We write $\overrightarrow{AB} + \overrightarrow{BC} = \overrightarrow{AC}$, which defines vector addition.
This does *not* mean $AB + BC = AC$, where AB, BC and AC are merely lengths.

To add vectors **u** and **v**, draw $\overrightarrow{AB}$ and $\overrightarrow{BC}$
nose–to–tail. Then $\overrightarrow{AC}$ represents **u** + **v**.

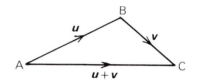

2 In the diagram, $\overrightarrow{PQ} + \overrightarrow{QR} = \overrightarrow{PR}$, so $\overrightarrow{PR}$ represents **u** + **v**.

Using components, $\begin{pmatrix} 3 \\ 3 \end{pmatrix} + \begin{pmatrix} 1 \\ -3 \end{pmatrix} = \begin{pmatrix} 4 \\ 0 \end{pmatrix}$.

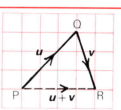

Exercise 5

1 Write down each vector addition, using: **a** directed line segments **b** components.

(i) 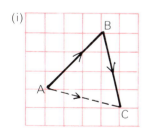 (ii) 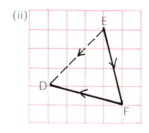 (iii)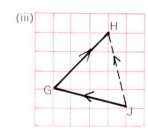

2 Copy each figure, and insert arrows so that $\overrightarrow{PQ} + \overrightarrow{QR} = \overrightarrow{PR}$.

a 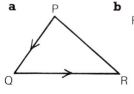 **b** 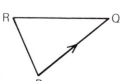 **c** 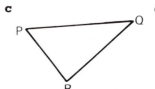 **d**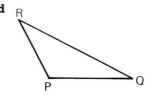

3 Plot the points K(1, 2), L(4, 5), M(5, 0).

  a Draw a vector triangle to show $\overrightarrow{KL} + \overrightarrow{LM} = \overrightarrow{KM}$.

  b Write each vector in component form, and check that $\overrightarrow{KL} + \overrightarrow{LM} = \overrightarrow{KM}$.

4

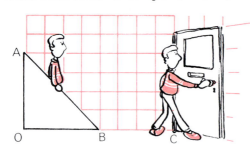

Martin comes down the stairs and along the hall to the front door. Copy and complete:

  (i) $\overrightarrow{AB} + \ldots = \ldots$

  (ii) $\begin{pmatrix} 4 \\ \ldots \end{pmatrix} + \begin{pmatrix} \\ \end{pmatrix} = \begin{pmatrix} \\ \end{pmatrix}$.

5 Find m and n: **a** $\begin{pmatrix} 3 \\ 2 \end{pmatrix} + \begin{pmatrix} -1 \\ -2 \end{pmatrix} = \begin{pmatrix} m \\ n \end{pmatrix}$ **b** $\begin{pmatrix} 1 \\ 2 \end{pmatrix} + \begin{pmatrix} -4 \\ n \end{pmatrix} = \begin{pmatrix} m \\ 0 \end{pmatrix}$.

6 Combine all of these vectors into one by:

  a drawing them nose–to–tail   **b** adding components

$\overrightarrow{AB} = \begin{pmatrix} 3 \\ 3 \end{pmatrix}$, $\overrightarrow{BC} = \begin{pmatrix} 2 \\ 0 \end{pmatrix}$, $\overrightarrow{CD} = \begin{pmatrix} 1 \\ 4 \end{pmatrix}$, $\overrightarrow{DE} = \begin{pmatrix} 2 \\ -3 \end{pmatrix}$, $\overrightarrow{EF} = \begin{pmatrix} 3 \\ 0 \end{pmatrix}$, $\overrightarrow{FG} = \begin{pmatrix} -3 \\ -2 \end{pmatrix}$, $\overrightarrow{GH} = \begin{pmatrix} -2 \\ 0 \end{pmatrix}$, $\overrightarrow{HI} = \begin{pmatrix} -6 \\ -2 \end{pmatrix}$.

7 The Jolly Harvester has to sail from A to B, past Seagull Island. The captain can steer a course to the North of the island in two straight lines, changing direction at a grid crossing point—or South of the island in the same way.

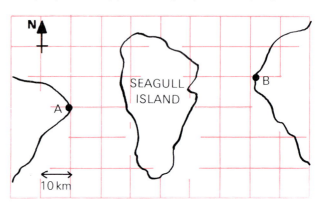

  a Write down the components of the two: (i) Northern legs   (ii) Southern legs.

  b Add the two vectors for each route, and compare your answers.

  c Which is the shorter route?

8 A helicopter flies from A to B over Seagull Island. What is the vector for its journey? Calculate its length, correct to 1 decimal place.

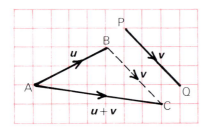

A problem! How to add $\overrightarrow{AB}$ and $\overrightarrow{PQ}$.

Remember—vector addition is *nose–to–tail*.

Solution Slide $\overrightarrow{PQ}$ to position $\overrightarrow{BC}$, so $\overrightarrow{BC} = \overrightarrow{PQ}$.

Then $\overrightarrow{AB} + \overrightarrow{PQ}$

      $= \overrightarrow{AB} + \overrightarrow{BC}$ *Check* $\begin{pmatrix} 4 \\ 2 \end{pmatrix} + \begin{pmatrix} 3 \\ -3 \end{pmatrix} = \begin{pmatrix} 7 \\ -1 \end{pmatrix}$.

      $= \overrightarrow{AC}$.

Note In science and engineering the sum of two vector quantities such as force, velocity and acceleration, is usually called their *resultant*.

=== *Exercise 5B* ===

1 a Use squared paper or tracing paper to draw:

   (i) $\boldsymbol{u}+\boldsymbol{v}$　(ii) $(\boldsymbol{u}+\boldsymbol{v})+\boldsymbol{w}$
(iii) $\boldsymbol{p}+\boldsymbol{q}$　(iv) $(\boldsymbol{p}+\boldsymbol{q})+\boldsymbol{r}$.

b Check, by adding components.

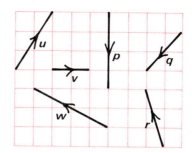

2

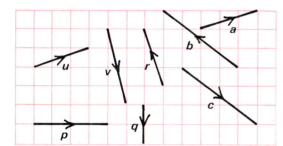

Which vector in the diagram is equal to each sum of vectors?

   (i) $\boldsymbol{u}+\boldsymbol{v}$
  (ii) $\boldsymbol{v}+\boldsymbol{u}$　Comment.
 (iii) $(\boldsymbol{p}+\boldsymbol{q})+\boldsymbol{r}$
 (iv) $\boldsymbol{p}+(\boldsymbol{q}+\boldsymbol{r})$　Comment.

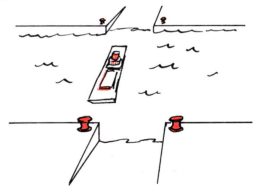

3 If there were no current, the ferry Faithful would cross the river with velocity $\overrightarrow{AB}$. But the current has velocity $\overrightarrow{CD}$.

a Use your ruler, or tracing paper, to find which of the numbered points the ferry arrives at on the North bank.

b Describe, or sketch, the ferry's velocity on the return trip so that the resultant velocity takes it to A.

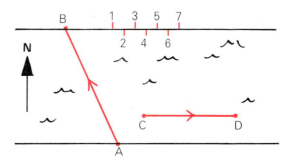

4

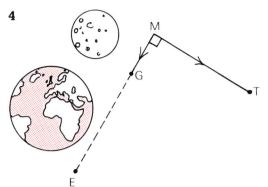

On its own, the Moon would travel from M to T in a given time. But the Earth's gravitational force pulls the Moon through the displacement $\overrightarrow{MG}$.

a Use tracing paper to find the resultant displacement $\overrightarrow{MT}+\overrightarrow{MG}$. Check that the Moon lies on a circle, centre E, radius EM.

b Repeat the process until the Moon is back at M.

c Approximately how long does the Moon take to go once round the Earth?

5 The directed line segments represent the forces on the aircraft in flight.

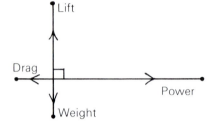

a By measuring and drawing, or tracing, find the resultant force for: (power+drag)+ (lift+weight).

b Is the plane moving forwards and upwards?

THE ZERO VECTOR

9 am: Pam's car is at the parking meter.

11 am: Her car is still there—or has it been away, and returned to the same place? The traffic warden would like to know!

In either case, its displacement is zero; an example of the zero vector.

> The zero vector has magnitude zero, and no defined direction.
>
> $\boldsymbol{0} = \begin{pmatrix} 0 \\ 0 \end{pmatrix}$.

Exercise 6

1 $\boldsymbol{p} = \begin{pmatrix} 7 \\ 3 \end{pmatrix}$, $\boldsymbol{q} = \begin{pmatrix} -2 \\ 5 \end{pmatrix}$, $\boldsymbol{r} = \begin{pmatrix} -4 \\ -2 \end{pmatrix}$. Write down vectors which give the zero vector when added to:

(i) $\boldsymbol{p}$ (ii) $\boldsymbol{q}$ (iii) $\boldsymbol{r}$ (iv) $\boldsymbol{p}+\boldsymbol{q}$ (v) $\boldsymbol{q}+\boldsymbol{r}$.

2 Find p and q: **a** $\begin{pmatrix} 3 \\ -4 \end{pmatrix} + \begin{pmatrix} p \\ q \end{pmatrix} = \begin{pmatrix} 0 \\ 0 \end{pmatrix}$ **b** $\begin{pmatrix} 5 \\ q \end{pmatrix} + \begin{pmatrix} p \\ -7 \end{pmatrix} = \begin{pmatrix} 0 \\ 0 \end{pmatrix}$.

3 a Find, in component form:
  (i) $\boldsymbol{u} + \boldsymbol{v}$ (ii) $\boldsymbol{p} + \boldsymbol{q}$
 b Which vector in the diagram added to $\boldsymbol{s}$
   gives the zero vector?
 c Use components to find:
  (i) $\boldsymbol{u} + \boldsymbol{s}$ (ii) $\boldsymbol{v} + \boldsymbol{r}$ (iii) $\boldsymbol{u} + \boldsymbol{s} + \boldsymbol{v} + \boldsymbol{r}$.

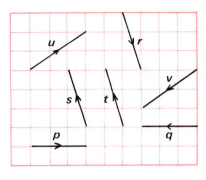

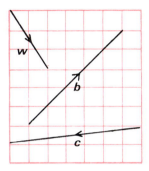

4

Torquil is battling against wind and current to pick up his lobster pots. The velocities of the wind ($\boldsymbol{w}$), current ($\boldsymbol{c}$), and boat in still water ($\boldsymbol{b}$) are shown in the diagram.
 a Construct the resultant vector $\boldsymbol{b} + \boldsymbol{c} + \boldsymbol{w}$ on squared paper.
 b Explain what is happening.

THE NEGATIVE OF A VECTOR

Can you solve the equation $x + 4 = 0$?
No problem, $x = -4$ ('negative 4').

Thinking in the same way for $\boldsymbol{u} + \boldsymbol{v} = \boldsymbol{0}$, $\boldsymbol{u} = -\boldsymbol{v}$, the negative of $\boldsymbol{v}$.

Example $\begin{pmatrix} 2 \\ 3 \end{pmatrix} + \begin{pmatrix} -2 \\ -3 \end{pmatrix} = \boldsymbol{0}$, so $\begin{pmatrix} -2 \\ -3 \end{pmatrix}$ or $-\begin{pmatrix} 2 \\ 3 \end{pmatrix}$ is the negative of $\begin{pmatrix} 2 \\ 3 \end{pmatrix}$.

=========== *Exercise 7* ===========

1 Write down the negative of each vector:
 a $\begin{pmatrix} 1 \\ 2 \end{pmatrix}$ **b** $\begin{pmatrix} -5 \\ -5 \end{pmatrix}$ **c** $\begin{pmatrix} 0 \\ 4 \end{pmatrix}$ **d** $\begin{pmatrix} -3 \\ 1 \end{pmatrix}$ **e** $\begin{pmatrix} 2 \\ -2 \end{pmatrix}$.

2

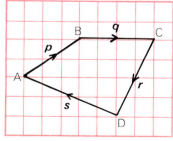

 a On squared paper draw vectors:
   (i) $-\boldsymbol{p}$ (ii) $-\boldsymbol{q}$ (iii) $-\boldsymbol{r}$ (iv) $-\boldsymbol{s}$.
 b Write each vector and its negative in components.

3 a Check that $\boldsymbol{u}+\boldsymbol{v}=-\boldsymbol{w}$.

  b In the same way, write down a vector for:

  (i) $\boldsymbol{w}+\boldsymbol{u}$ (ii) $\boldsymbol{v}+\boldsymbol{w}$ (iii) $\boldsymbol{u}+(\boldsymbol{v}+\boldsymbol{w})$.

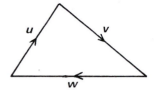

4

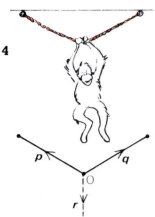

Ally the ape hangs quite still, because the resultant of the forces in the ropes balances his weight, that is $\boldsymbol{p}+\boldsymbol{q}=-\boldsymbol{r}$. The angles at O are equal, and $|\boldsymbol{p}|=|\boldsymbol{q}|$.

  a Draw a vector 'triangle of forces' to show $\boldsymbol{p}+\boldsymbol{q}=-\boldsymbol{r}$.

  b Write down a vector for: (i) $\boldsymbol{p}+\boldsymbol{r}$ (ii) $\boldsymbol{q}+\boldsymbol{r}$.

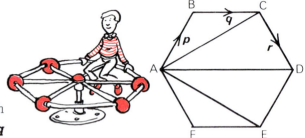

5 ABCDEF is a regular hexagon in which $\overrightarrow{AB}$, $\overrightarrow{BC}$ and $\overrightarrow{CD}$ represent vectors $\boldsymbol{p}$, $\boldsymbol{q}$ and $\boldsymbol{r}$.

  a Express in terms of $\boldsymbol{p}$, $\boldsymbol{q}$, $\boldsymbol{r}$:

  (i) $\overrightarrow{AF}$ (ii) $\overrightarrow{EF}$ (iii) $\overrightarrow{AC}$

  (iv) $\overrightarrow{AD}$ (v) $\overrightarrow{AE}$.

  b (i) Find $\overrightarrow{AB}+\overrightarrow{BC}+\overrightarrow{CD}+\overrightarrow{DE}+\overrightarrow{EF}+\overrightarrow{FA}$.

  (ii) What is the connection between $\overrightarrow{AB}+\overrightarrow{BC}+\overrightarrow{CD}$ and $\overrightarrow{DE}+\overrightarrow{EF}+\overrightarrow{FA}$?

In which of the following is: (i) $\boldsymbol{u}=-\boldsymbol{v}$ (ii) $|\boldsymbol{u}|<|\boldsymbol{v}|$ (iii) $|\boldsymbol{u}|>|\boldsymbol{v}|$?

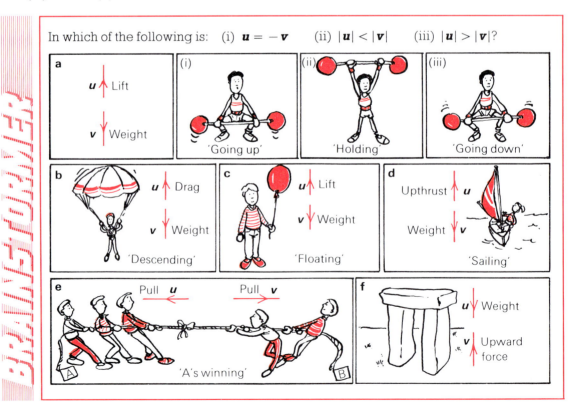

SUBTRACTION OF VECTORS

Subtract 5 from 7? No problem, add its negative. $7-5 = 7+(-5) = 2$.

Thinking in the same way, $\boldsymbol{u} - \boldsymbol{v} = \boldsymbol{u} + (-\boldsymbol{v})$.

Example

a $\boldsymbol{p} = \begin{pmatrix} 3 \\ 4 \end{pmatrix}$, $\boldsymbol{q} = \begin{pmatrix} 2 \\ -3 \end{pmatrix}$ so $-\boldsymbol{q} = \begin{pmatrix} -2 \\ 3 \end{pmatrix}$

$\boldsymbol{p} - \boldsymbol{q} = \boldsymbol{p} + (-\boldsymbol{q}) = \begin{pmatrix} 3 \\ 4 \end{pmatrix} + \begin{pmatrix} -2 \\ 3 \end{pmatrix} = \begin{pmatrix} 1 \\ 7 \end{pmatrix}$.

b Look at the diagram. To find $\boldsymbol{p} - \boldsymbol{q}$:
 (i) reverse the direction of $\boldsymbol{q}$
 (ii) use the nose–to–tail addition of vectors.

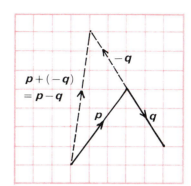

=================== *Exercise 8* ===================

1

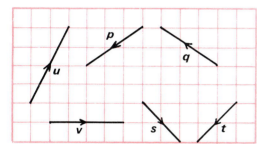

a Write down the components of each vector in the diagram.
  Find the components of:
b (i) $\boldsymbol{p} - \boldsymbol{q}$ (ii) $\boldsymbol{q} - \boldsymbol{p}$
c (i) $\boldsymbol{u} - \boldsymbol{v}$ (ii) $\boldsymbol{v} - \boldsymbol{u}$
d (i) $\boldsymbol{s} - \boldsymbol{t}$ (ii) $\boldsymbol{t} - \boldsymbol{s}$.

2 On plain paper, sketch $\boldsymbol{p}$, $\boldsymbol{q}$, $-\boldsymbol{q}$ and $\boldsymbol{p} - \boldsymbol{q}$ for these:

(i)   (ii)   (iii)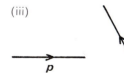

3 $\boldsymbol{m} = \begin{pmatrix} 4 \\ 1 \end{pmatrix}$ and $\boldsymbol{n} = \begin{pmatrix} 1 \\ 3 \end{pmatrix}$.

 a Find in component form: (i) $\boldsymbol{m} + \boldsymbol{n}$ (ii) $\boldsymbol{m} - \boldsymbol{n}$.
 b Show $\boldsymbol{m}$, $\boldsymbol{n}$, $\boldsymbol{m} + \boldsymbol{n}$ and $\boldsymbol{m} - \boldsymbol{n}$ in one diagram on squared paper.

4 PQRS is a parallelogram.
 a Name two directed line segments representing: (i) $\boldsymbol{u}$ (ii) $\boldsymbol{v}$.
 b Name a directed line segment representing:
  (i) $\boldsymbol{u} + \boldsymbol{v}$ (ii) $\boldsymbol{u} - \boldsymbol{v}$.

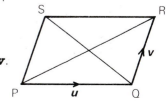

INTRODUCING VECTORS

5

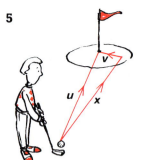

Andy is on the putting green. The slope will take the ball off line with displacement **v**.

a Write a vector equation with **x** as subject.

b Andy is 3 m South of the hole, and the slope will take the ball 50 cm West. To sink his putt, Andy will have to aim slightly to the right. Calculate the angle between this line and the direct line to the hole.

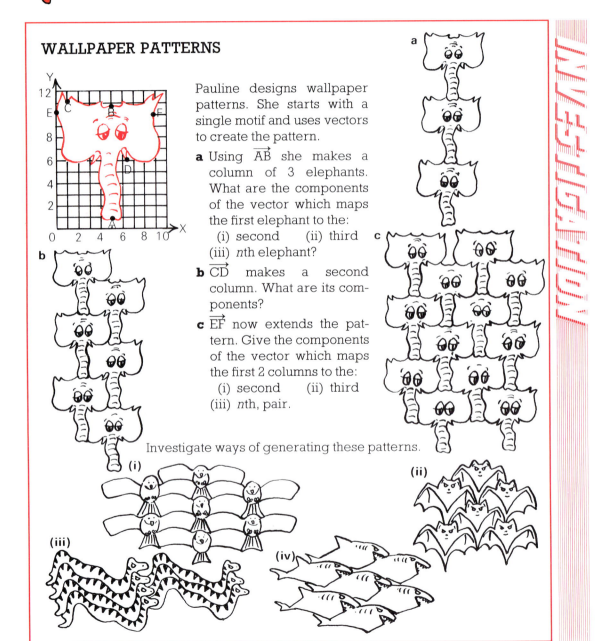

WALLPAPER PATTERNS

Pauline designs wallpaper patterns. She starts with a single motif and uses vectors to create the pattern.

a Using $\overrightarrow{AB}$ she makes a column of 3 elephants. What are the components of the vector which maps the first elephant to the:
   (i) second   (ii) third
   (iii) nth elephant?

b $\overrightarrow{CD}$ makes a second column. What are its components?

c $\overrightarrow{EF}$ now extends the pattern. Give the components of the vector which maps the first 2 columns to the:
   (i) second   (ii) third
   (iii) nth, pair.

Investigate ways of generating these patterns.

(i)   **(ii)**

(iii)   **(iv)**

MULTIPLICATION OF A VECTOR BY A NUMBER

The wallpaper patterns suggest that adding a vector $1, 2, 3, \ldots, n$ times has the effect of multiplying its components by $1, 2, 3, \ldots, n$. For example, in the diagram, $\boldsymbol{v} + \boldsymbol{v} + \boldsymbol{v}$ produces a vector with the same direction and three times the magnitude.

$$\boldsymbol{v} + \boldsymbol{v} + \boldsymbol{v} = \begin{pmatrix} 1 \\ 2 \end{pmatrix} + \begin{pmatrix} 1 \\ 2 \end{pmatrix} + \begin{pmatrix} 1 \\ 2 \end{pmatrix} = 3\begin{pmatrix} 1 \\ 2 \end{pmatrix} = \begin{pmatrix} 3 \\ 6 \end{pmatrix}.$$

So $\boldsymbol{v} + \boldsymbol{v} + \boldsymbol{v} = 3\boldsymbol{v}$.

> If $\boldsymbol{v} = \begin{pmatrix} a \\ b \end{pmatrix}$, then $k\boldsymbol{v} = k\begin{pmatrix} a \\ b \end{pmatrix} = \begin{pmatrix} ka \\ kb \end{pmatrix}$.

Example 1

If $\boldsymbol{v} = \begin{pmatrix} 1 \\ 2 \end{pmatrix}$, $-2\boldsymbol{v} = -2\begin{pmatrix} 1 \\ 2 \end{pmatrix} = \begin{pmatrix} -2 \\ -4 \end{pmatrix}$.

$-2\boldsymbol{v}$ has twice the magnitude of $\boldsymbol{v}$, and the opposite direction.

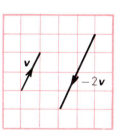

Example 2

If $\boldsymbol{u} = \begin{pmatrix} 3 \\ 12 \end{pmatrix}$, $\frac{1}{3}\boldsymbol{u} = \frac{1}{3}\begin{pmatrix} 3 \\ 12 \end{pmatrix} = \begin{pmatrix} 1 \\ 4 \end{pmatrix}$.

$\frac{1}{3}\boldsymbol{u}$ has $\frac{1}{3}$ magnitude of $\boldsymbol{u}$, and the same direction.

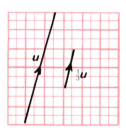

=== *Exercise 9* ===

1

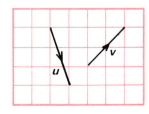

a On squared paper, draw:
(i) $3\boldsymbol{u}$ (ii) $3\boldsymbol{v}$ (iii) $3\boldsymbol{u} + 3\boldsymbol{v}$ (iv) $\boldsymbol{u} + \boldsymbol{v}$ (v) $3(\boldsymbol{u} + \boldsymbol{v})$.
b Express each in component form.
c Comment on the results.

2 a Write $\boldsymbol{p}$, $\boldsymbol{q}$, $\boldsymbol{r}$, $\boldsymbol{s}$ in component form.

b Write the following vectors in components, and draw them on squared paper:

(i) $4\boldsymbol{p}$ (ii) $\frac{1}{2}\boldsymbol{q}$ (iii) $-2\boldsymbol{r}$

(iv) $\frac{2}{3}\boldsymbol{s}$ (v) $-\frac{3}{5}\boldsymbol{r}$.

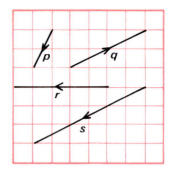

3 Calculate the magnitude of each vector.

For example, $5\begin{pmatrix}6\\8\end{pmatrix} = \begin{pmatrix}30\\40\end{pmatrix}$. The magnitude $= \sqrt{(30^2 + 40^2)} = 50$.

a $2\begin{pmatrix}3\\4\end{pmatrix}$ **b** $5\begin{pmatrix}9\\-12\end{pmatrix}$ **c** $\begin{pmatrix}-8\\-15\end{pmatrix}$.

4 Solve these vector equations for $\boldsymbol{x}$:

a $\boldsymbol{x} - 2\boldsymbol{u} = \boldsymbol{v}$ **b** $2\boldsymbol{v} - \boldsymbol{x} = \boldsymbol{u}$ **c** $2\boldsymbol{x} + \boldsymbol{u} = \boldsymbol{v} - 4\boldsymbol{x}$.

5 Find the components of $\boldsymbol{x}$ in each solution in question **4**, given $\boldsymbol{u} = \begin{pmatrix}2\\-5\end{pmatrix}$ and $\boldsymbol{v} = \begin{pmatrix}4\\2\end{pmatrix}$.

6 John is out rowing. The current is running at 10 m/s in a northerly direction, and the wind's velocity is 16 m/s from the West.

a What speed and direction has John to row, just to stand still?

b Repeat **a** if the speed of the current doubles and the wind speed halves.

7 Investigate the relationship between $\boldsymbol{u}$ and $k\boldsymbol{u}$ for: **a** $k > 0$ **b** $k < 0$.

CHECK-UP ON **INTRODUCING VECTORS**

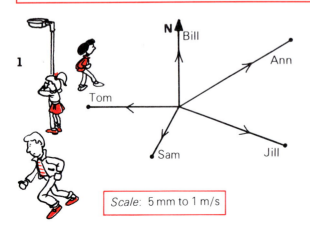

1

Scale: 5 mm to 1 m/s

The diagram shows five velocity vectors of children running away for 'Hide–and–Seek'. Bill runs North at 4 m/s. Measure the magnitude and direction of each vector with your ruler and protractor.

2 $p = \begin{pmatrix} 2 \\ 3 \end{pmatrix}$ and $q = \begin{pmatrix} -3 \\ 4 \end{pmatrix}$. Calculate: **a** $p + q$ **b** $p - q$ **c** $2p - 3q$.

3 a Make a vector equation with $\overrightarrow{AB}$, $\overrightarrow{BC}$ and $\overrightarrow{AC}$.
   b Express it in component form.

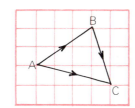

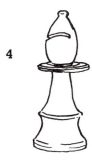

4

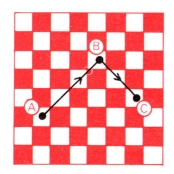

In chess, the Bishop moves diagonally only.
   a Write down the components of the Bishop's moves from:
   (i) A to B   (ii) B to C
   b Calculate $|\overrightarrow{AC}|$.

5 ABCD is given a displacement which takes
   A(1, 2) to A′(6, 3), B to B′, C to C′, D to D′.

   a What are the components of the displacement?
   b Calculate the magnitude of the displacement.
   c Write down the coordinates of B′, C′ and D′.

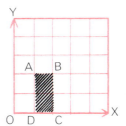

6 a Make a sketch of the diagram.
   b On your sketch, draw:
   (i) $u + v$ (ii) $u - v$
   c (i) $|u + v| = |u - v|$. Why?   (ii) $u + v \neq u - v$. Why not?

7 Every metre forward, Rosalind's trolley slips 30 cm sideways to the right. Calculate the magnitude and direction (compared to straight ahead) of the resultant displacement of the trolley.

8 A Knight in the middle of the chess board can visit 8 squares (see Exercise 3, question **4**). Investigate its possible displacements in the course of *two* moves.

REVISION 1A

1 Write down the next two numbers that could be in these sequences, and state the rule you use:

    a 1, 3, 6, 10, 15, . . .   **b** 100, 93, 97, 90, 94, . . .   **c** 1, 4, 9, 16, ...

2 Simplify:

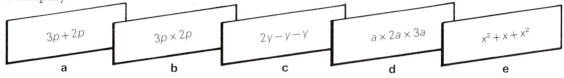

$3p + 2p$	$3p \times 2p$	$2y - y - y$	$a \times 2a \times 3a$	$x^2 + x + x^2$
a	**b**	**c**	**d**	**e**

3

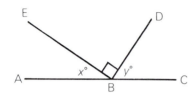

ABC is a straight line.

    a Write down an equation involving x and y.

    b If $x = 35$, find y.

    c If $x > 30$, what can you say about y?

    d If $x = y$, what can you say about x and y?

4

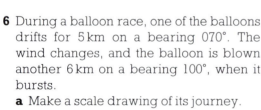

39 cm

June saves 5p coins in this tube. Each coin is 1·5 mm thick. How much money will she have when the tube is full?

5 $x = -1$, $y = -4$ and $z = 2$. Calculate the values of:

    a x^2 **b** xyz **c** $2y^2$ **d** $2z - xy$ **e** $\dfrac{yz}{x}$ **f** $\dfrac{x + y + z}{3}$.

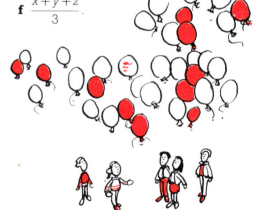

6 During a balloon race, one of the balloons drifts for 5 km on a bearing 070°. The wind changes, and the balloon is blown another 6 km on a bearing 100°, when it bursts.

    a Make a scale drawing of its journey.

    b Measure the distance and bearing of the balloon's bursting point from its starting point.

7 A table-top football pitch is a scaled down version of the real thing. The scale is 1 : 45. Calculate:

    a the breadth of the model if the breadth of the real pitch is 63 m

    b the length of the real pitch if the model is 190 cm long.

8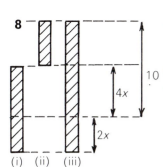

(i) (ii) (iii)

a Find an expression for the length of each of the three planks. The lengths are in metres.

The total length of the three planks is 28 metres.

b Make an equation in x and solve it.

c Write down the lengths of the planks.

9 a Prove that $\triangle$s ADE and ABC are similar.

b Write down the scale factor which reduces $\triangle$ABC to $\triangle$ADE.

c Write down a ratio equal to $\dfrac{x}{x+4}$.

d Calculate x (the lengths are in cm).

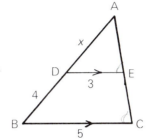

10 Action High School has 90 fifteen-year old pupils who travel by bus—30 from Midton, 40 from Easton and 20 from Weston. Use this information to draw:

a a pie chart   **b** a bar graph.

11 Factorise:

a $6m-9n$ **b** m^2-mn **c** m^2-n^2 **d** m^2-2m+1 **e** m^2+m-6.

12

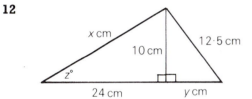

Calculate: **a** x

b y

c z.

13 Mrs Watt's electricity bill drops into her letterbox.

Domestic charge:

17·15p for each of the first 52 units
5·85p for each remaining unit.

Off–peak:

5·85p for each of first 78 units
3·65p for each remaining unit.
How much has she to pay?

	Meter readings	
	Present	Previous
Domestic	10436	10178
Off–peak	02041	01118

(VAT is zero–rated)

14 Solve:

a $6x-3=2x-11$ **b** $3(y+2)>12$ **c** $t^2-3t-4=0$.

REVISION 1B

1 Write down the next two numbers that could be in these sequences, and state the rule you use:

a $2, -4, 8, -16, \ldots$ **b** $1, 2, 3, 5, 8, 13, \ldots$ **c** $2, 3, 5, 8, 12, \ldots$.

2 Solve:

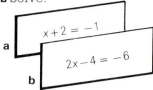

a $x + 2 = -1$

b $2x - 4 = -6$

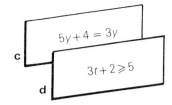

c $5y + 4 = 3y$

d $3t + 2 \geqslant 5$

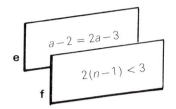

e $a - 2 = 2a - 3$

f $2(n - 1) < 3$

3 a Draw x and y–axes on squared paper from -6 to 6, and draw the triangle with vertices A(6, 5), B(3, 4), C(3, 2).

b Draw $\triangle$A′B′C′, the image of $\triangle$ABC under reflection in the y–axis.

c Draw $\triangle$A″B″C″, the image of $\triangle$A′B′C′ in the x–axis. How could you obtain $\triangle$A″B″C″ directly from $\triangle$ABC?

4 The time taken by Mr Hughes to drive between two service stations on a motorway is inversely proportional to his average speed. He takes 45 minutes at an average speed of 48 km/h. Calculate:

a his average speed when the journey takes 50 minutes

b his time for the journey at an average speed of 90 km/h.

5 Copy the table, and fill in each box.

x	1	3			10
y	5	2	3		10
	6	5	7	9	
	25	4	9	81	

(No negative numbers allowed.)

6

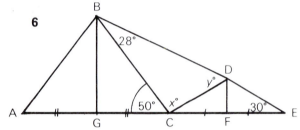

ACE is a straight line, and $\triangle$s ABC and DCE are isosceles.

a Why must GB be parallel to FD?

b Copy the diagram, and find x and y.

7 A rectangular tank measures 3 m by 2·5 m by 1·2 m. Calculate the number of litres of water it will hold when it is half-full.

8 Multiply out: **a** $(a + b)^2$ **b** $(a - b)^2$ **c** $(3x + 1)(x + 1)$ **d** $(4d - 1)(3d + 2)$

e $(2t + 3)^2$ **f** $\left(x - \dfrac{1}{x}\right)^2$ **g** $(a + b)(a - b + 1)$ **h** $(k + 2)^3$.

9

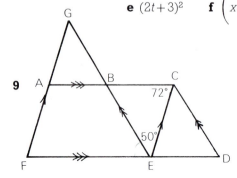

a Name an angle:

(i) corresponding to $\angle$ GAB

(ii) alternate to $\angle$ CBE

(iii) vertically opposite $\angle$ GBC.

b Copy the diagram and fill in the sizes of all the angles.

10 The entries in Claire Brown's salary slip are the same each month. Her income tax allowances total £3845, and she is taxed at 27p in the £ on her taxable income. Calculate the missing items in her salary slip.

CLAIRE BROWN		EMPLOYEE 0138	TAX CODE 384L	MONTH 4
BASIC PAY 1235·25	OVERTIME —	COMMISSION —	BONUS 55·25	GROSS PAY
INCOME TAX	SUPERANNUATION —	NATIONAL INSURANCE 82·75	OTHER DEDUCTIONS 6·50	TOTAL DEDUCTIONS
				NET PAY

11

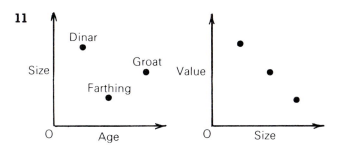

a Label the second graph with groat, Dinar and farthing.
b Which coin has least value?
c Draw a graph of age against value for the three coins.

12 AB is a diameter of the circle. Calculate:
    a BC   **b** AC, correct to the nearest cm
    c the area of △ABC, correct to the nearest cm².

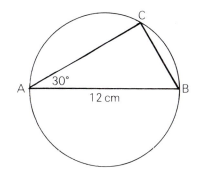

13 Mr and Mrs Taylor wish to insure their house and contents. The house is valued at £45 000, their valuables at £3500 and other contents at £12 000. They insure the house for 1½ times its value at £0·19 per £100, valuables at £1·25 per £100 and other contents at £0·68 per £100. Calculate their annual premium.

14 Find an expression for:
    a the total volume of the three boxes (the lengths are in metres)
    b the total exposed surface area of the boxes when stacked as shown.

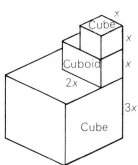

REVISION 2A

1

TOP PRINTS

Number of exposures	Price
20	£3
24	£3·36
36	£5·20

Which would you buy, and why?

2 Solve: **a** $9x - 70 = 11$ **b** $3y = y - 2$ **c** $x - 7 = 87 - x$ **d** $4n + 2 = 6n - 12$.

3 Copy the diagrams, and fill in the sizes of as many angles as you can.

a
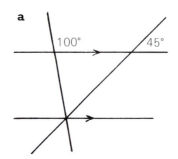

b
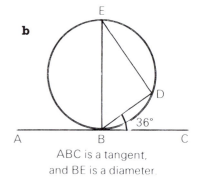

ABC is a tangent, and BE is a diameter.

4 Neptune is 4500 million km from the sun, and the eccentricity of its orbit is 0·00855. Write both of these numbers in standard form, $a \times 10^n$.

5 Mr Goodlad's shopping list. How much?

250 g cheese at £2.52 per kg
2½ dozen eggs at 45p per ½ dozen
½ lb chopped pork at 78p per lb
1¾ lb sirloin steak at £2.60 a lb

6 Copy the table, and fill in each box.

a	3	−2	4	−3		−1
b	2	−1	1	1	3	
	5	−3			5	
		1	−1			0

7 On squared paper, draw the line from O to A(7, 7). Draw the triangle with vertices (1, 1), (1, 6) and (3, 6). Draw the image of this triangle under reflection in OA (so that OA becomes an axis of symmetry of the complete figure). Write down the coordinates of its vertices.

8 Calculate the perimeter and area of this metal plate. Make a sketch first. All the angles are right angles.

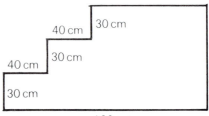

9

$2x - 1$ cm

$8x - 7$ cm

The three short straws have the same total length as the long straw. Calculate the length of each kind.

10 Happy Hire charge £25 a day for car hire, plus 30p a mile, with a minimum charge of 20 miles. After 100 miles, the rate is reduced to 15p a mile. Follow the flowchart to find the hire charge for a mileage of:

a 18 miles   **b** 30 miles   **c** 90 miles   **d** 200 miles.

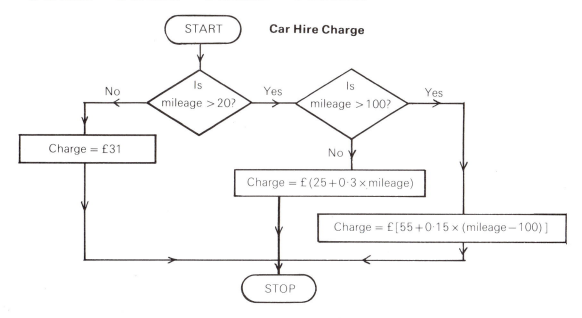

11 Moya is a window–dresser in Tog's Boutique. Her basic working week is 37 hours at £4·50 an hour. Overtime is paid at 'time-and-a-third'. Calculate her earnings in a week when she works: **a** 40 hours   **b** 43 hours.

12

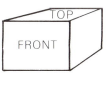

2 cm — TOP

3 cm — FRONT

4 cm

Bill has started to draw the net of a box.

a Write down the length, breadth and height of the box.

b Draw two possible nets for the box.

13 a Make an inequation for each picture, and solve it.

b Use the two solutions to find all possible numbers of weights in the bags.

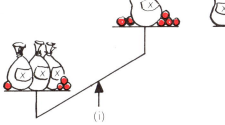

(i)

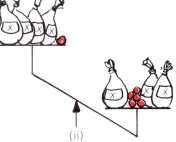

(ii)

Each bag contains x weights of 1 kg. The weights are 1 kg each also.

14 Investigate whether $y \propto x$, $y \propto \dfrac{1}{x}$, or neither, for each table:

a

x	5	25	45	75
y	3	15	27	45

b

x	20	40	60	100
y	2·5	5	7·5	12·5

c

x	2	3	4	6	8
y	30	20	15	10	7·5

15 Find the value of: **a** $9^{1/2}$ **b** $8^{-1/3}$ **c** 5^{-2} **d** $(3^2)^{-1}$ **e** 11^0 **f** $\sqrt[3]{8^2}$.

REVISION 2B

1 The Healthy Glow chain of food stores employs check–out staff, all paid at the same rate. One store needs 11 check–out staff, with a wage bill of £935 per week.
  a Calculate the wage bill at another branch which needs 9 check–out staff.
  b East Port branch has a weekly wage bill of £1105. How many check–out staff does it employ?

2 Solve for t: **a** $-8 + 2t = -6$ **b** $\dfrac{3t}{2} = -6$ **c** $2t + a = 3b$ **d** $p = q + \sqrt{t}$.

3 Measurements are made on a base-line AB to find the height of the steeple CD. Calculate, correct to the nearest metre:
  a BC **b** CD.

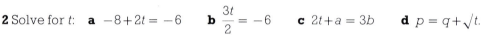

4 Eight hungry people are waiting to take the lift to the restaurant on the top floor. Their weights (kg) are: 82, 52, 55, 60, 73, 68, 48 and 61.

How can they be 'lifted' in two trips without exceeding the maximum load?

5 Work out all possible sums, differences, products and quotients of pairs of terms on the cards.

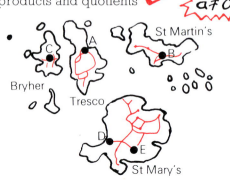

6 The map shows some of the Scilly Isles. The distance between places A and B is 4 km.
  a Make a scaled ruler to use with the map.
  b Use your ruler to find these distances, to the nearest km:
  (i) AC (ii) AD (iii) AE.

7 Tom drives for 15 minutes at an average speed of 48 km/h, then for half an hour at an average speed of 72 km/h.

a Calculate his average speed for the whole journey.

b The return journey takes 45 minutes. He drives for the first 25 minutes at an average speed of 60 km/h. Find his average speed for the rest of the journey.

8   Find the cost of 1 jar of jam and the cost of 1 roll. (Remember to check your answer.)

Total cost £2·20   Total cost £3·15

9 A committee of Euromathematicians decide to decimalise angle measure: 100 new degrees, called 'decigrees', to a revolution (complete turn).

a How many old degrees are in a decigree?

b How could a Euromathematician define:

(i) a right angle   (ii) an acute angle   (iii) an obtuse angle   (iv) a reflex angle?

c Draw a N–S–E–W diagram and mark the eight main compass points as bearings from North in decigrees.

10 Sunshine frequency in 1987 at Dunedin weather station:

No. of hours sunshine (to the nearest hour)	0	1	2	3	4	5	6	7	8	9	10	11	12	13	14
Number of days	20	25	19	32	29	39	45	40	33	24	18	20	12	7	2

a Make a frequency table, and calculate the mean number of hours sunshine.

b Draw a histogram. What is the modal number of hours sunshine?

11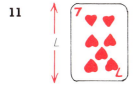

a Show that when one card is placed at right angles on another the area A cm² of the shape formed is $A = 2LB - B^2$.

b Each card is 8·8 cm by 5·8 cm. Calculate A.

c Change the subject of the formula to L, and use your formula to calculate L when $A = 84$ and $B = 6$.

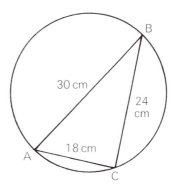

12 a Prove that AB is a diameter of the circle.

b Calculate the perpendicular distance from the centre of the circle to:

(i) BC   (ii) AC.

13  Boxes of Fruitos are made in two sizes. The small boxes are similar to the large ones.

   a What is the scale factor for the enlargement from small to large size?

   b A small box is 6 cm high. Calculate the height of a large box.

   c The front of a small box has area 27 cm². Calculate the area of the front of a large box.

   d The volume of a small box is 54 cm³. Calculate the volume of a large box.

14 Simplify: **a** $\dfrac{x-2}{3}+\dfrac{2x}{5}$ **b** $\dfrac{y+1}{2}-\dfrac{y-2}{3}$ **c** $\dfrac{3}{p+q}-\dfrac{2}{p-q}$ **d** $\dfrac{1}{t^2-4}+\dfrac{1}{t-2}$.

15 The cosine rule for $\triangle ABC$ is $a^2 = b^2+c^2-2bc\cos A$.

   a Make cos A the subject.

   b Calculate the largest angle of the triangle with sides 55 mm, 75 mm and 95 mm long.

REVISION 3A

1 Calculate:

   a $40-(3\times11)$ **b** $\dfrac{9+(6\div2)}{(4\times1)-1}$ **c** $7+(\tfrac{1}{2}\text{ of }1\tfrac{1}{2})$

   d $\tfrac{1}{7}$ as a decimal, rounded to 5 significant figures.

2 Multiply out, and simplify where possible.

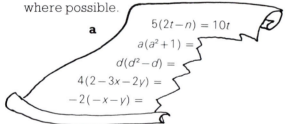

   a
$$5(2t-n) = 10t$$
$$a(a^2+1) =$$
$$d(d^2-d) =$$
$$4(2-3x-2y) =$$
$$-2(-x-y) =$$

   b
$$-(2-k) =$$
$$5-3(x+1) = 5-3x-$$
$$4x-2(1-x) =$$
$$-6(-a+b-1) =$$
$$1-(1-x)-(1+x) =$$

3 a On squared paper draw the kite O, A(1, 2), B(4, 0), C(1, −2).

   b Draw a tiling of kites around it.

   c What can you say about the sum of the angles around each vertex in the tiling?

4 How many cartons of Juicy can be packed in a box 1·2 m by 0·8 m by 0·45 m?

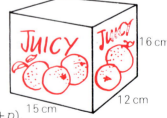

16 cm

15 cm

12 cm

5 Solve these inequations:

   a $2(2p-3) < 4$ **b** $6p+3(1-p) \geqslant -6$ **c** $4(p-2)-p \leqslant -(16+p)$.

6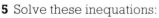

50 cm

A game has a square board of side 50 cm, with a circular track of width 4 cm reaching the sides. Using 3·14 for π (or the π key), calculate:

   a the circumference of the outside circle

   b the area of the track.

7
BRITBANK
Interest—
7% p.a.

How much interest would £1800 earn in each bank in 8 months?

MONEYBANK
Interest—
$7\frac{1}{2}$% p.a.

8 The ladder is 1 metre longer than the distance it reaches up the wall. Make an equation, and solve it to find the distance the ladder reaches up the wall.

4 m

9

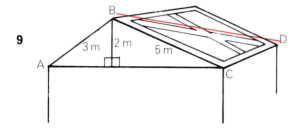

The factory roof is asymmetrical (the South-facing slope is larger to catch more sunlight).
a Calculate all the angles of △ABC, correct to 0·1°.
b ∠CBD = 55° and ∠BCD = 90°. Calculate the length of BD, correct to 0·1 m.

10

1 card on top

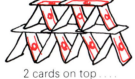

2 cards on top . . . .

a Copy and complete:

No. of cards on top (T)	1	2	3	4
Total no. of horizontal cards (H)	3			
Total no. of cards (C)	13			

b Find a formula for: (i) H in terms of T (ii) C in terms of T.
c Is it possible to find a formula for C in terms of H? Investigate.

11

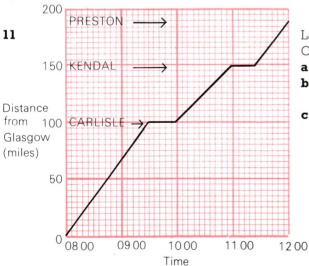

Distance from Glasgow (miles)

Time

Laurie drives a van from Glasgow to Preston. Calculate:
a the length of each stop he makes
b his average speed on each of the three parts of his journey
c his average speed for the whole journey
(i) including stops (ii) excluding stops.

12 A piece of elastic is pinned at B and C.
BC = 30 cm. It is pulled out to A, so that
△ABC is isosceles.
a How long is AC:
(i) before stretching   (ii) when $d = 20$?
b The elastic is stretched from its initial position
until its length (CA + AB) is increased by a
factor of 2·6. Calculate d now.

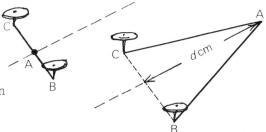

13 Solve one of these pairs of equations by drawing graphs, one by substitution of x or y, and
one by eliminating x or y (by adding or subtracting suitable equations):

a $\left.\begin{array}{l} 2y+x = 12 \\ 3y-x = 18 \end{array}\right\}$ **b** $\left.\begin{array}{l} y = x \\ y = 2x+11 \end{array}\right\}$ **c** $\begin{array}{l} x+2y = 12 \\ 2x-y = -11 \end{array}$.

14 Which number less than 70 has the largest number of distinct factors. Investigate.

REVISION 3B

1 A meal at the Ritz Restaurant costs Karen £12·25, including VAT at 15%. What does the meal
cost without VAT?

2 Solve: **a** $3(2x-1) = x+2$ **b** $2(1-2x) = 2-3(x-1)$ **c** $2(x+1)-3(x-1) = 6$.

3 ACB is the cross–section of a rain gutter
at the edge of the roof. It is part of a circle
of radius 6 cm, and its depth is 4 cm.
Calculate the width AB of the gutter, in
simplest surd (square root) form.

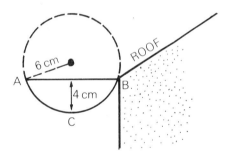

4 Which conditions does
each point satisfy?

Conditions	Points
(i) $x \geqslant 2$ and $y < 3$	A(2, 2)
(ii) $1 < y \leqslant 3$	B(1, −1)
(iii) $x < 2$ or $y > -1$	C(0, 0)
	D(3, 0).

5 Two ships leave port P at the same time. One
sails 40 km on course 040° to A. The other
sails 50 km on course 097° to B. Calculate (to
the nearest metre and degree):   **a** AB
b ∠PAB   **c** the bearing of B from A.

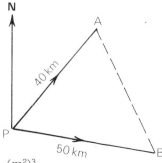

6 Simplify:

a $\dfrac{x^4 \times x^6}{x^2}$ **b** $y^{1/3}(y^{2/3}+y^{-1/3})$ **c** $(t^{-1/2})^4$ **d** $\dfrac{(m^2)^3}{(mn)^2}$.

7 The rates of income tax in 1987 were:

Taxable income (£)	Rate (%)
1 to 17 900	27
17 901 to 20 400	40
20 401 to 25 400	45
25 401 to 33 300	50

How much tax was paid on taxable incomes of:
a £8000
b £18 000
c £28 000?

8 a

SQUARE THEN ADD 1
Output?

b A function f is defined by $f(x) = (2x - 1)^2$.
Calculate: (i) $f(1)$ (ii) $f(-2)$.
c For another function g, $g(x) = \sin x°$.
Calculate:
(i) $g(30)$ (ii) $g(123)$, to 1 decimal place.

9

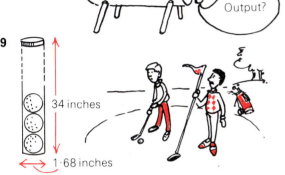

34 inches

1·68 inches

Dimhope golf balls are packed in tubes.
Calculate:
a the number of golf balls per tube
b the percentage of unused space in the tube.

10 You can write 24 as the product of several pairs of factors: 1×24, 2×12, and so on; also -1×-24, -2×-12, ...
  a Write down all the pairs you can find. Plot them on squared paper like this: $(1, 24)$, $(2, 12)$ . . ., and draw a graph.
  b What is the equation of the graph? What shape is the graph—a parabola or a hyperbola?

11 Find the solution of each equation to 3 significant figures.
  a Using the quadratic formula: (i) $x^2 + x - 1 = 0$ (ii) $2x^2 - x - 5 = 0$ (iii) $x(x - 2) = 9$.
  b Using the step-by-step method: (i) $2x^2 - x - 2 = 0$ (the root between 1 and 2)
  (ii) $x^2 + 4x + 2 = 0$ (the root between -1 and 0).

12 A 'rep-tile' is a shape which tiles to form larger versions of itself.

For example, the square is a rep-tile since four squares tile as shown to make a larger square.

  a Show that a 2 by 1 rectangle is a rep-tile.
  b What is the least number of equilateral triangles needed to show that it is a rep-tile?
  c Try to show that this sphinx shape is a rep-tile. It is made from equilateral triangles.

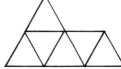

13 In a Fourth Form of 80 pupils, 50 take art, 35 music and 20 craft and design. 15 pupils take art and music only, 6 take art with craft and design only, but none take music with craft and design only. Using a Venn diagram, find how many pupils take all three subjects.

REVISION 4A

1 a From 15 May to 23 August (including both dates) the car ferry Glen Sannox makes four trips daily to Arran. How many trips altogether?

b A train leaving London Euston at 22 50 arrives at Glasgow Central at 07 35, having been delayed at Carlisle from 04 50 to 05 25. Without this delay, how long would the journey have taken?

2 Solve: **a** $6(x+1)-2x = x+4$ **b** $\dfrac{x-2}{2}+\dfrac{x}{3}=4$ **c** $2x^2-5x-3=0$.

3 A triangular plot of ground ABC has each side 100 m long. A TV aerial has to be situated on the plot, subject to these conditions:
  (i) It must be at least 25 m from each side of the plot.
  (ii) It must be within 55 m of A.
    Make a diagram, and show clearly the region where the mast can go.

4 a List all the prime numbers less than 100.
  b How many are: (i) 1 more than a multiple of 6
                          (ii) 1 less than a multiple of 6
                          (iii) in neither (i) nor (ii)?
  c Calculate the percentage of each type.

5 Fab Fashions increased its profits in a year from £140 000 to £185 000.
  a Calculate the percentage increase in profits.
  b As a result, employees were given a 15% discount on purchases. How much would they pay for the following?

(i)

SCARF
£3·60

(ii)

COAT
£45·40

(iii)

HAT
£29·68

6 Tom flies his model plane in a circle at the end of a wire. The tension in the wire varies directly as the square of the plane's speed and inversely as the radius of the circle. The tension is 8 newtons for a speed of 8 m/s and a radius of 6 m. Calculate the tension for a speed of 10 m/s and a radius of 9 m.

7 A bottle of diameter 8 cm sits in a rack. The points of contact P and R are 3 cm from A. Calculate:
  a AT   **b** $\angle$PAR.

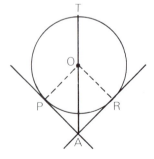

8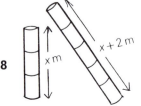

The sections making up the pipes are all the same length.
  a For each pipe, write down the length of a section in terms of x.
  b Make an equation, solve it and write down the length of each pipe.

9 a Calculate: (i) $2\frac{3}{4}+1\frac{3}{8}$ (ii) $4\frac{1}{5}-2\frac{2}{5}$ (iii) $3\frac{3}{4}\times1\frac{1}{5}$ (iv) $2\frac{5}{8}\div1\frac{3}{4}$.

b How many $\frac{3}{4}$ kg bags of salt can be filled from a 60 kg supply?

c Two slots in the barrier of a ripple tank are $\frac{3}{8}$ mm and $\frac{7}{16}$ mm wide. A third slot is needed with width halfway between the other two. Calculate the width of this third slot.

10 Solve for x:

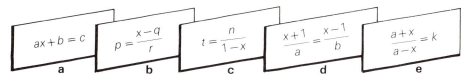

$$ax+b=c \qquad p=\frac{x-q}{r} \qquad t=\frac{n}{1-x} \qquad \frac{x+1}{a}=\frac{x-1}{b} \qquad \frac{a+x}{a-x}=k$$

   a   **b**   **c**   **d**   **e**

11 a Prove: (i) $\triangle$ADC is right–angled.
           (ii) $\triangle$s ABD and CBD are similar.

b Calculate: (i) BD
           (ii) the area of each of the three triangles.

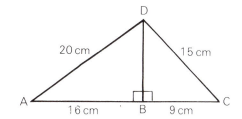

12 Savewise Building Society offers 7% p.a. interest (tax paid) on its accounts. A National Savings Bank Investment account pays 11% p.a. interest, but you may have to pay tax on this at 27p in the £.

a For an investment of £850, which is better for:

(i) Ann, who does not pay tax   (ii) Alison who pays tax at the basic rate?

b Calculate the compound interest on £850 invested with Savewise for 3 years.

13 The height (h metres) of a rocket after t seconds is given by the formula $h=20t-4t^2$.

a Copy and complete the table of values.

b Draw the graph of h against t.

c How long is the rocket in the air?

d Solve the equation: (i) $20t-4t^2=0$
                (ii) $20t-4t^2=21$.

e At what times is the rocket 12 m above the ground?

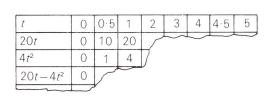

t	0	0·5	1	2	3	4	4·5	5
$20t$	0	10	20					
$4t^2$	0	1	4					
$20t-4t^2$	0							

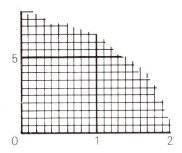

14

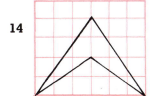

Boomerangs are shaped like arrowheads.

a Calculate: (i) the area of the kite (in squares)
               (ii) the sizes of the angles in the kite.

b By tiling, find the smallest rectangle from which eight arrowheads can be made.

REVISION 4B

1 a The Hill family changed £1200 to Swiss francs at 2·40 francs to the £ for their holiday. How many francs did they receive?

b On the ferry home they changed their remaining 173 Swiss francs to £s at 2·36 to the £. How much were they given?

2 Solve:

a $\dfrac{1}{x}+\dfrac{1}{2}=0$ **b** $\dfrac{1}{2}(2y-1)=-3$ **c** $\dfrac{x-1}{2}+\dfrac{x+1}{3}=4$ **d** $\dfrac{x}{2}+\dfrac{2}{x}=2.$

3 This chess board stretches on and on . . .
A square is defined by the coordinates of its bottom left-hand corner. The Knight moves from $(0,0)$ to $(1,2)$ to $(2,4)$, and so on in the same way.

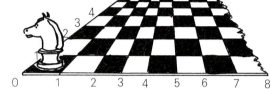

a List the first six squares he visits.
b If he lands on $(x,20)$ and $(17,y)$, find x and y.
c (x,y) is one of his squares. Write an equation in x and y.

4 a The length and breadth of a rectangle are each increased ·by 10%. Calculate the percentage increase in the area of the rectangle.

b $2\frac{1}{2}\%$ of a firm's workforce are paid off. This left a workforce of 897. How many lost their jobs?

5

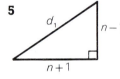 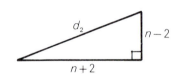 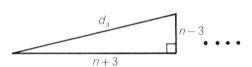

a Find expressions for: (i) d_1^2, d_2^2 and d_3^2 (ii) $d_2^2-d_1^2$ and $d_3^2-d_2^2$.
b Continuing the sequence in (ii), write down an expression for $d_4^2-d_3^2$, and try to discover a formula for $d_{k+1}^2-d_k^2$.
c If n is a positive integer, how does the sequence end?

6 Kwikheat Co makes central heating systems. It carries out tests to compare the efficiency of different systems.

Time (min)		0	10	20	30	40	50
Temperature (°C)	Type A	5	6	8	12	17	24
	Type B	5	8	13	17	20	21

The tests shows how long it takes for two systems to raise the temperature of a room.
a On the same diagram draw a graph of temperature against time for each set of results.
b After how long are the temperatures equal?

7 Simplify:

a $\dfrac{5n+15}{3n+9}$ **b** $\dfrac{2x+6}{x^2+2x-3}$ **c** $\dfrac{a^2-b^2}{3a-3b}$ **d** $\dfrac{3y^2-12}{2y^2+3y-2}$

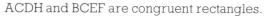

8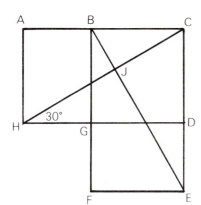

ACDH and BCEF are congruent rectangles.
a Copy the diagram, and by filling in angles show that BE and HC cross at right angles.
b Explain why the result is true for any size of $\angle$ CHD.
c Prove that HG = FG.
d Name three points which are vertices of an isosceles triangle with HC as one side.

9 Jim buys canvas to make a wigwam for his son Ian. He wants the wigwam to be 2 m in diameter and 3 m high.
a What area of canvas is needed?
b Calculate the volume of the tent.

10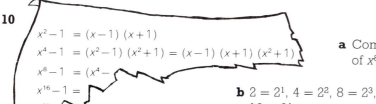

$$x^2 - 1 = (x-1)(x+1)$$
$$x^4 - 1 = (x^2-1)(x^2+1) = (x-1)(x+1)(x^2+1)$$
$$x^8 - 1 = (x^4 -$$
$$x^{16} - 1 =$$
$$x^{32}$$

a Complete the factorisation of $x^8 - 1$, $x^{16} - 1$ and $x^{32} - 1$.

b $2 = 2^1$, $4 = 2^2$, $8 = 2^3$, $16 = 2^4$, ...
How many factors have: $x^2 - 1$, $x^4 - 1$, $x^8 - 1$, $x^{16} - 1$?

c How many factors has $x^{2^n} - 1$?

11 Results of a survey of hourly rates of pay (£) of shop assistants in the High Street:

1·60	1·85	1·50	2·12	1·98	2·60	2·48	1·58	1·75	1·94
2·66	2·35	2·60	1·95	2·10	1·65	1·78	2·26	2·34	2·00
1·85	1·90	2·12	1·84	2·45	2·50	1·86	2·00	2·35	2·05

a Construct a frequency table, with class intervals 1·50–1·64, etc.
b Calculate the mean rate of pay, using the frequency table.
c Add a cumulative frequency column, and draw a cumulative frequency curve.
d Use this curve to estimate the median rate.

12 Solve:
a $2x^2 - 8 = 0$ **b** $(2x-1)^2 - 9 = 0$ **c** $2x^2 + 3x - 4 = 0$, correct to 2 decimal places.

13 a On the same diagram *sketch* the graphs of $y = \sin x°$, $y = \sin 2x°$ and $y = \sin\frac{1}{2}x°$ for $0 \leqslant x \leqslant 360$. What is the period of each?
b Draw the graphs of $y = \sin x°$ and $y = \sin\frac{1}{2}x°$ for $90 \leqslant x \leqslant 180$. Use them to help you to solve the equation $\sin x° = \sin\frac{1}{2}x°$ for $90 \leqslant x \leqslant 180$.

14 Write in a simpler surd (square root) form:
a $\sqrt{20}$ **b** $4\sqrt{5} - 2\sqrt{5}$ **c** $\sqrt{6}(\sqrt{2} + \sqrt{3})$ **d** $(\sqrt{3} - \sqrt{2})^2$ **e** $\dfrac{3}{2\sqrt{6}}$.

Number families

a Check each line in the table with your calculator.

b Write down: (i) the next three members of the family
(ii) the tenth member
(iii) the nth member.

c The first member of Family 2 is $5^2 + 12^2 = \ldots$
Copy and complete this.

d Write down the next two members.

e Family 3 begins $7^2 + \ldots = 25^2$. Copy and complete.

f Copy and complete the first member of Family 4: $9^2 + \ldots = \ldots$

g What is the link between all of these families and a theorem in geometry?

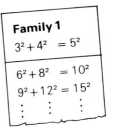

Family 1

$3^2 + 4^2 = 5^2$

$6^2 + 8^2 = 10^2$

$9^2 + 12^2 = 15^2$

$\vdots \quad \vdots \qquad \vdots$

How to win a magazine competition

Win £1000! All you have to do is to choose 8 items you would find most useful, and list them in order 1 to 8.

How can you be sure of submitting a correct entry?

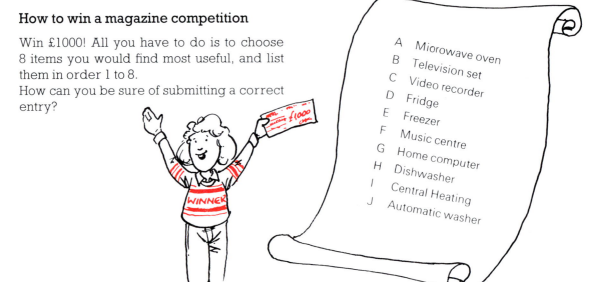

A Microwave oven
B Television set
C Video recorder
D Fridge
E Freezer
F Music centre
G Home computer
H Dishwasher
I Central Heating
J Automatic washer

Think it out! Two items A and B can be listed in 2 ways, AB and BA.

In how many ways can three items, A, B, C be listed, two at a time? Make lists to find out.

There's a quicker way! Think of filling the first place, then the second, and so on. Work out a system.

Then find how many entries you would have to send in to the competition to be sure of winning. A magazine costs 50p. Any snags?

A fixture fix

Phil's in a fix. He is Secretary of the Bluebell Badminton League. There are six teams, each of which plays the others once. The teams play one game a week, on Wednesdays at 7 pm. What is the least number of weeks needed to complete the fixtures? Make out a fixture list.

Leaning towers

1 There are many leaning towers in Italy. The most famous one is at Pisa. Calculate its angle of tilt.

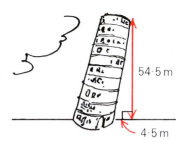

54·5 m

4·5 m

2 At some stage this empty box will topple:

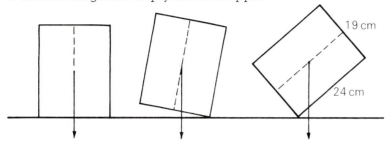

19 cm

24 cm

a Sketch the box when it is on the point of toppling over.
b Calculate its angle of tilt at this point.
c Check your answer by tilting your mathematics book about one corner, and measuring the angle of tilt with a protractor.

3 A crate is $3x$ m by $2x$ m by x m.
Calculate the angles at which it will topple about each edge.

On course

Sarah is piloting a small aircraft from A to B. She takes off at 1000 hours, and after 25 minutes she finds that she is 10° off course at X, less than half way to B. She wonders whether to turn 10° or 20° clockwise to get back on course, and what to do then. Draw a diagram, and give instructions which would solve her problem and ensure that she reaches B.

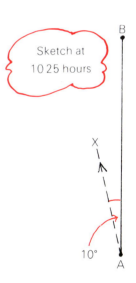

B

Sketch at
1025 hours

X

10°
A

Next–door neighbours

Bandit Builders have a piece of land for nine houses. Each house is built on a square plot with a fence right round it. One houseowner will have a special problem, but that is not for you to worry about.

A	E	B
H	I	F
D	G	C

How many next–door neighbours (who share a fence) will each houseowner have?
Bandit Builders buy more land, and decide to build in square blocks of 4, 9, 16, 25 and 36 houses. Copy and complete:

Number of houses		4	9	16	25	36
Number of next–door neighbours	2		4			
	3		4			
	4		1			

Investigate the patterns of numbers in the rows, and then continue the table up to 144 houses. If a field is broken up into n^2 plots how many owners have:
a 2   **b** 3   **c** 4 next–door neighbours?
Investigate how you would modify the results for square plots in a rectangular field, m plots long and n plots broad.

A pyramid puzzle

HI–IQ Toy Co. want a design for sets of pyramids that can fit together to make cubes of side 16 cm. Sketch one of the pyramids, and calculate its dimensions. If you have time you could check by making the cube.

Order from chaos

It is often necessary to arrange lists in increasing order. Here's a program that does it for you. (A, B, C are all different).
1 Try the program for A = 3, B = 2, C = 1, writing down the results at stages (i), (ii) and (iii).
2 Try it for data 30, 50, 40.
3 The computer understands > to mean 'is after . . . , alphabetically'. For example, Bert > Alan. Use the flowchart to sort the data: Peter, Viv, Bill.

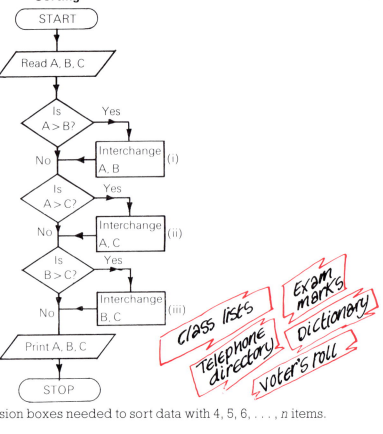

Sorting

Investigate the number of decision boxes needed to sort data with 4, 5, 6, . . . , n items.

A statistical survey

Choose a topic:

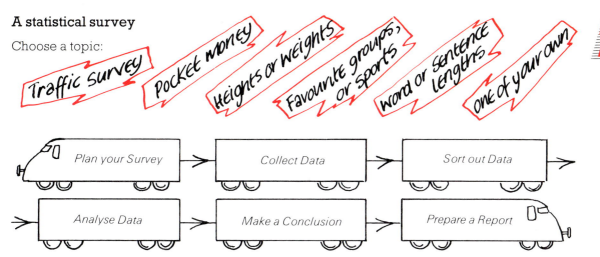

Remember—tables and graphs are always useful.

Frequency tables
and tallies

Pictographs

Bar graphs

Pie charts

Line graphs

Mean

Median

Mode

Range

Percentages

Could your survey be used, improved, extended?

A magic square

1 Copy and complete the Magic Square. Check that the numbers in each row, column and diagonal add up to the same total.

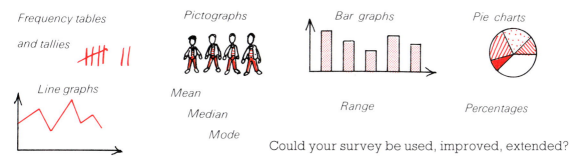

$a-b$	$a+b-c$	$a+c$
$a+b+c$	a	$a-b-c$
$a-c$	$a-b+c$	$a+b$

Values
$a = 5$
$b = 4$
$c = 1$

1	8	

2 Try it again for $a = 2$, $b = -1$, $c = 3$.   **3** Explain how it works.

4 For all entries to be different, are there any restrictions on a, b, c?

5 If you would like a challenge, find a 3×3 magic square whose entries are all powers of 2 and whose *products* of the entries in each row, column and diagonal are all equal.

Ruler and compasses only

Frank says that it is possible to draw a circle whose area is the sum of the areas of these two circles, using ruler and compasses only.

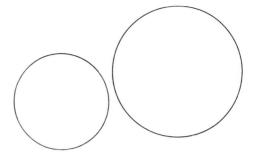

Later he found that it was possible to draw a circle whose area was the difference between the areas of the two circles. Can you manage either, or both? Remember Pythagoras?

A house of cards

Mike is building a house of cards. For 1 storey he uses 2 cards, and makes 1 triangle of 1 unit.

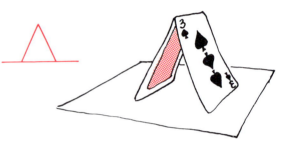

For 2 storeys he uses 5 *more* cards, a total of 7. He makes 4 triangles of 1 unit and 1 triangle of 4 units.

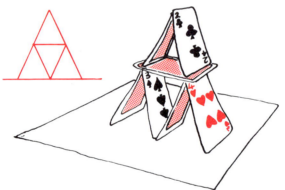

1 For 3 storeys, how many:
   a more cards
   b cards altogether
   c 1–unit triangles
   d 4–unit triangles
   e 9–unit triangles?

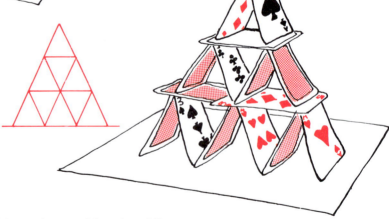

2 Arrange all the data in a table, and extend it to 4 and 5 storeys.

3 Find formulae for:
   (i) the number of 1–unit triangles in n storeys
   (ii) the number of extra cards needed for the nth storey.

4 How many storeys can Mike build with 2 packs of cards (52 in each)?

Table tennis balls

Table tennis balls (diameter 4 cm) are to be sold in sets of six. Packem Box Co. have been asked to design containers for them. Investigate this design problem.

Consider the best shapes for economy of material or space, for packing containers together, for display, and so on. You'll have to calculate some areas and volumes. Then construct a container and test it.

The snooker player

Johnny's snookered. He has to hit the black, but another ball is in the way. Make a larger copy of the diagram, and assuming that the angle of rebound is equal to the angle of approach of the ball to the cushion, try to find the point X.

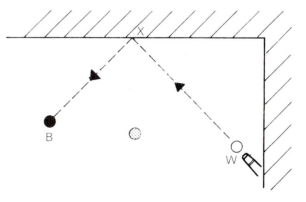

A mathematical solution to Johnny's problem

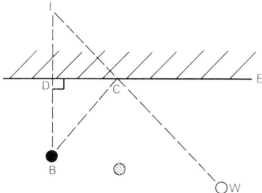

b Calculate:
  (i) the angle of approach, ∠WCE
  (ii) the distance CE.
c This time there is another ball at X (top diagram), so Johhny has to make a '2-cushion escape'. Investigate.

a I is the image of the black ball in the cushion. Copy the diagram, and explain why ∠WCE = ∠BCD.

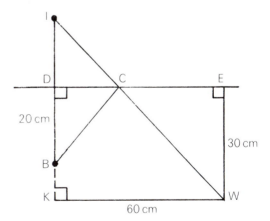

Football scores

List all possible half-time scores in the Rovers v United game. Try to arrange them in a methodical way.

Repeat this for two or three different full-time scores.

If the final score is 'Rovers m United n' can you make a formula in terms of m and n for the number of possible half-time scores?

Planning a project

Choose one of the following, or a project of your own. Think out the details and costs, in order, and prepare a report.

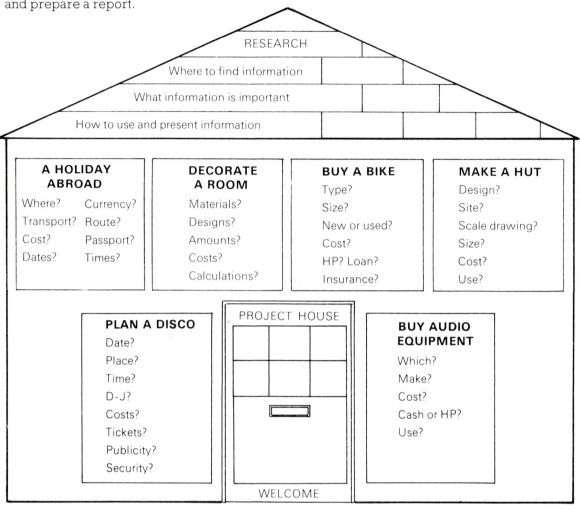

RESEARCH

Where to find information

What information is important

How to use and present information

A HOLIDAY ABROAD

Where?	Currency?
Transport?	Route?
Cost?	Passport?
Dates?	Times?

DECORATE A ROOM

Materials?

Designs?

Amounts?

Costs?

Calculations?

BUY A BIKE

Type?

Size?

New or used?

Cost?

HP? Loan?

Insurance?

MAKE A HUT

Design?

Site?

Scale drawing?

Size?

Cost?

Use?

PLAN A DISCO

Date?

Place?

Time?

D-J?

Costs?

Tickets?

Publicity?

Security?

PROJECT HOUSE

BUY AUDIO EQUIPMENT

Which?

Make?

Cost?

Cash or HP?

Use?

WELCOME

Shop codes

A multiple store gives its branches 2–digit code numbers based on 0 and 1. For example, High Street Store is coded 01, and Low Street Store is coded 00. List all possible code numbers. How many are there?

Another, larger firm, uses 0, 1 and 2 for 2–digit code numbers. List all possible code numbers. Copy and complete the table for 2–digit codes.

Number of digits used	2	3	4	5	6	10	. . .	n
Total number of codes	4						. . .	

Add more rows to your table for the total number of 3–digit, 4–digit . . . , n–digit code numbers.

The King's problem

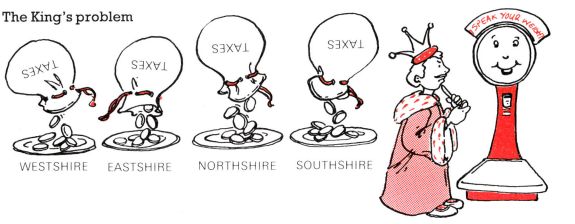

WESTSHIRE   EASTSHIRE   NORTHSHIRE   SOUTHSHIRE

A powerful King receives his taxes in gold coins from his four regions. Each coin weighs 100 g, but a spy tells him that one of his Regional Governors is cheating him by melting down the coins and mixing in a lighter, worthless metal. Each coin from this region weighs 90 g. Unfortunately the King has only a very, very old weighing machine and only one old penny left to make the machine speak. Using only one weighing, how can the King find out which Governor is cheating him?

How to 'prove' that $2 = 1$

$$a = b$$
$$a^2 = ab \dots\dots\text{Why?}$$
$$a^2 - b^2 = ab - b^2 \dots\text{Why?}$$
$$(a-b)(a+b) = b(a-b) \dots\text{Why?}$$
$$a + b = b \dots\dots\text{Why?}$$
Let $a = b = 1$
Then $2 = 1$ What has gone wrong?

Traffic lights survey

Plan, and carry out, a survey at a set of traffic lights. In a report include a sketch of the road junction, light times, number of cars waiting, time for pedestrians to cross, effect on traffic flow, suggestions for improvements, and so on.

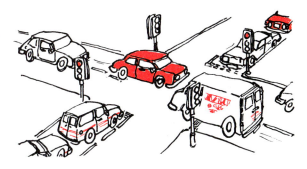

Expansions

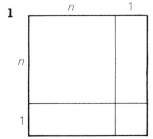

Copy the squares and rectangles. By filling in areas, check that $(n+1)^2 = n^2 + 1 + 2n$, the expansion for $(n+1)^2$.

2 Here is an exploded view of a cube with length of side $n+1$ units. Write down the volume of each lettered part, and find an expansion for $(n+1)^3$.

Why would this method not work for $(n+1)^4$? Can you work out the expansion of $(n+1)^4$?

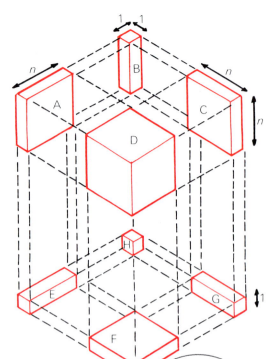

Friendly circles

The radius of each of the three congruent, touching circles is 10 cm. The small circle in the middle has centre O and radius r.

1 a Why is $\angle ABO = 90°$.
   b What size is $\angle AOB$?
   c Calculate OA, and write down the value of r.

2 Repeat **1** for four congruent circles.

3 How many circles would there have to be for the inside circle to have radius 10 cm also?

Polydiags and the 'Mystic Rose'

1 a Sketch the square, pentagon and hexagon. How many diagonals has each?

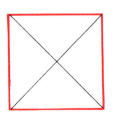 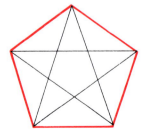 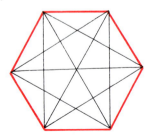

   b Continue the investigation for polygons with $7, 8, 9, \ldots$ sides.
   c Show that the formula for the number (N) of diagonals in an n–sided polygon is $N = \frac{1}{2}n(n-3)$.

2 To make the Mystic Rose you draw all possible lines from every vertex. That means all the sides *and* diagonals.

  a How many lines are needed for polygons with 3, 4, 5 . . . sides.
  b Find a formula for the number of lines in an *n*–sided Mystic Rose.

3 How could the first formula have helped you with the second?

A calculator search

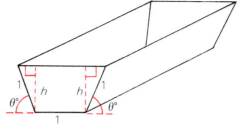

The feeding–trough has a cross–section in the shape of a trapezium. The base and sloping edges of the trough are 1 m long. Each edge makes an angle $\theta°$ with the ground.

  a Find an expression for *h* in terms of $\sin \theta°$, $\cos \theta°$ or $\tan \theta°$.
  b Show that the area of the cross–section is $(1 + \cos \theta°) \sin \theta°$.
  c Use your calculator to investigate the areas given by different values of θ.
  d Find the value of θ that maximizes the capacity of the trough.

Make–believe

The table is up-to-date. Can you work out the results of all the games played so far?

SCOTLAND 2   ENGLAND 0

— **GOALS** —

	Played	Won	Drawn	Lost	For	Against	Points
Scotland	3	3	0	0	8	0	6
N. Ireland	2	1	0	1	4	2	2
Wales	2	0	1	1	0	4	1
England	3	0	1	2	0	6	1

Flower petals

You know how to fix the position of a point using *cartesian co-ordinates (x, y)*. Another way is to use *polar coordinates* $(r, \theta°)$. P is the point $(6·4, 20°)$. Copy and complete:

$\theta°$	0	10	20	30	40	45	50	60	70	80	90
$2\theta°$	0	20	40								
$r = 10 \sin 2\theta°$	0	3·4	6·4								

Plot the points $(r, \theta°)$ and draw the graph of $r = 10 \sin 2\theta°$ for $0 \leqslant \theta \leqslant 90$. Continue the table at 10° intervals until you can complete the four 'flower petals'.
Investigate the graph of $r = 10 \sin \frac{1}{2}\theta°$ for $\theta = 0, 30, 60, \ldots, 360$. You'll be surprised by the shape of this graph.

MRS M. SIME.